BASIC
ELECTRICAL
MEASUREMENTS

Steven Geczy

Mohawk College, Hamilton, Ontario

BASIC ELECTRICAL MEASUREMENTS

PRENTICE-HALL, INC., Englewood Cliffs, New Jersey 07632

Library of Congress Cataloging in Publication Data

GECZY, STEVEN, 1925–
    Basic electrical measurements.

      1. Electric measurements. I. Title.
TK275.G38 1984   621.37´4   83-13663
ISBN 0-13-060285-X

Printed in the United States of America

10 9 8 7 6 5 4 3 2 1

Editorial/production supervision and interior design by Paul Spencer
Cover design: Wanda Lubelska
Manufacturing buyer: Gordon Osbourne

ISBN 0-13-060285-X

Prentice-Hall International, Inc., *London*
Prentice-Hall of Australia Pty. Limited, *Sydney*
Editora Prentice-Hall do Brasil, Ltda., *Rio de Janeiro*
Prentice Hall Canada Inc., *Toronto*
Prentice-Hall of India Private Limited, *New Delhi*
Prentice-Hall of Japan, Inc., *Tokyo*
Prentice-Hall of Southeast Asia Pte. Ltd., *Singapore*
Whitehall Books Limited, *Wellington, New Zealand*

Contents

3   DC VOLTAGE MEASUREMENTS   53

4   AC VOLTAGE MEASUREMENTS   77

5   CURRENT MEASUREMENT   85

6   RESISTANCE MEASUREMENT   99

7   POTENTIOMETRIC CIRCUITS   123

APPENDIX   133

SOLUTIONS TO SELECTED PROBLEMS

Preface

The measurement of any variable requires knowledge of measurement units, measuring techniques, and some properties of the measured variable, but above all it requires *good judgment*. A measured value can be put to use only if the measurer is sure of the validity of his or her figures and is able to express them in proper form.

The aim of this book is to develop the ability to make good measurements as much as it is to impart the factual knowledge of the rules and methods. The principles outlined above are therefore never let out of sight. Thus the reader is enabled to acquire both the techniques and the judgment that are necessary for measurement of any given variable.

"Basic" in the title means that the book does not handle more material than students can comfortably digest in one semester; it does not imply simplistic treatment. For the topics covered, this book gives in-depth explanation of all significant details, treating the subjects more completely than other books of comparable aims.

There is no special prerequisite for this book beyond basic algebra, though some previous experience with basic electricity can be helpful. Explanations are kept tight, however, and repetitions are avoided, to make the book compact and easy to use for reference.

Steven Geczy

LIST OF SYMBOLS

A	ampere
I	current, overall current
I_m	PMMC full scale deflection current
R	resistance, overall resistance
R_a	adjustable resistor
R_b	zero balancing, compensating resistor
R_c	resistor for reference voltage adjustment
R_d	diode resistance
R_e	equivalent resistance
R_r	ranging resistor
R_{sh}	shunt resistor
R_x	unknown resistor
V	volt, voltage source
V_{ave}	average voltage
V_c	reference voltage
V_e	equivalent voltage
V_o	original measured voltage
V_p	peak voltage
V_{rms}	RMS voltage
Ω	ohm
Ⓥ	voltmeter
Ⓐ	ammeter
⊘	PMMC movement
↑	null detector

BASIC
ELECTRICAL
MEASUREMENTS

1

Principles of Measurement

1.1 SYSTEMS OF MEASUREMENT

Like the issuance of money, the establishment of systems of measures was historically the perquisite of sovereigns. Not until the advent of the Industrial Revolution did it become necessary to organize a system that was generally applicable and accepted. Two of these, the imperial system and the metric system, prevailed, gobbling up and superseding most other systems.

In the late 1920s (history was kind to forget the exact date) it became known that the yardstick, used as the primary standard for the imperial system, was shrinking. The yardstick made its transition from science into history when the inch was set at exactly 25.4 millimetres. Thus did the imperial system cease to exist, giving way to the now generally recognized metric system, which can be used either with traditional metric units or with translated traditional units.

The development of metric units did not go without a hitch. The metre is *not* exactly 1/40 000 000 part of the meridian of Paris, and the kilogram is *not* exactly the mass of 1 litre of 4°C water at the stated conditions; however, that did not prevent the establishment of recognized, copiable primary standards.

The translation of units (that is, the establishment of conversion factors) became the duty of legislative bodies. For example, His Majesty's Parliament established the gallon (imperial) as 4.546 092 litres. H.M. Canadian Parliament established it as 4.546 09 litres, and the U.S. Congress (still imperial) as 4.542 494 litres. These differences were not great enough to cause problems at the gas pump, but were enough to rule out the use of such translated units for science, research, and high technology. Because of these differences, units of this period should properly be called "U.K.," "Canadian," or "U.S." rather than "British" or "imperial."

There were problems with the metric system as well, the most prominent being the fact that it was based on weight (force) rather than mass, and the conditions under which the calorie should be measured were not agreed upon.

To stop the discrepancies and to arrive at a generally applicable system the SI system (Système International d'Unités) was created.

This system took over many of the metric units, discarded others, and created new ones. The SI system was officially adopted in 1960, but work on it went on until 1971.

Right now there is one (*and only one*) acceptable system of measurement for scientific use, the SI system. Old units of metric and imperial origin are still used, and they are harmless unless used for something they cannot do. [For example, the traditional unit for heat energy is the Btu. Which Btu? The international tables (1055.06 J), the mean (1055.87 J), the thermochemical (1054.35 J), the 60°F (1054.68 J), or the 39°F (1059.67 J)? Ask your air conditioner salesman.]

The SI system took over the traditionally used units for basic electrical quantities, but the units were redefined. Watch out for this when reading older texts about electrical measurements!

This book uses SI units throughout.

1.2 SI SYSTEM
The SI system consists of the definition of the basic units, the derivation of secondary units, and general rules about notation and the use of the units. No attempt is made in this book to present the SI system in its entirety. The only units that will be explained are those used in the text. For a description of the entire system, the student must obtain a document such as *The American National Standard on Metric Practice* (American National Standards Institute/The Institute of Electrical and Electronics Engineers, Standard 268-1982) or *Standard for Metric Practice* (American Society for Testing and Materials, E380-82)—in Canada, the *Canadian Metric Practice Guide* or Standard CSAZ 234.2-1973.

1.2.1 General Rules

1.2.1.1 Writing of Numbers

In writing down integers, the numbers must be grouped in sets of three digits. There must be a space between the groups: for example, 54 318 or 16 300 000. There should be no comma or any other separating mark between the numbers. The only exemption is thousands (four digits). These can be written as either 4 526 or 4526. Similarly, 3.4156 or 3.415 6 are both acceptable forms.

Decimals are separated by a decimal point. Writing on a blackboard, a point is sometimes not visible enough, and in practice a short comma may be used. This cannot lead to confusion, because any mark between two digits must be interpreted as the decimal sign.

When there are more than four digits after the decimal point, these digits must be grouped in threes and separated by a space, like the groups of integers: for example, 0.000 015 2 or 8 765 312.456 61.

1.2.1.2 Scientific Notation

There is a more convenient form of writing down large numbers than the usual long form. A decimal point is put after the first digit and the number is multiplied by the proper exponent of 10 to show the actual magnitude. For example, 165 000 000 becomes 1.65×10^8 and 98 650 becomes 9.865×10^4. Note that the exponent always indicates the number of places the decimal point was carried to the left.

The number of digits following the decimal point is determined by the accuracy of the number. Because of this, a zero or zeros can be carried at the end of a scientifically notated number. For example, 6.5400×10^7 indicates that the number is accurate to five significant digits (as will be explained in more detail later).

Scientific notation can also be used with numbers smaller than 1. For example, 0.000 067 52 becomes 6.752×10^{-5}. In this case, the value of the negative exponent equals the number of places the decimal point was carried to the *right*.

1.2.1.3   Engineering Notation

Engineering notation differs from scientific notation in the fact that only 3 and multiples of 3 (positive or negative) are used as exponents of 10. For example, in engineering notation 165 000 000 becomes 165×10^6 and 0.000 067 52 becomes 67.52×10^{-6}.

Several advantages of engineering notation are that it is easier to compare numbers of similar magnitude, and it is easier to convert numbers with units into or from numbers, using prefixes.

Multiplier prefixes must be given preference over both engineering and scientific notation. Prefixes are explained in Section 1.2.2.1.

1.2.1.4   Implied Accuracy

Numbers denote quantities which can be the result of a count or a measurement. A *count* is a *digital* quantity and as such is definite. For example, 236 head of sheep is exactly that—neither 235 nor 237 nor anything in between. A *measurement* is an *analog* quantity, subject to measurement inaccuracies and limitations. If a *measured quantity* (but not a number representing a count) is written down, that number represents a definite accuracy and precision.

Notice that the accuracy and precision of a number are not the same thing as the accuracy and precision of a measurement. This section discusses the accuracy and precision of numbers; the accuracy of measurements is discussed in Sections 1.3 and 1.4. Also, a written number implies a certain accuracy, but it does not guarantee that the measurement was actually taken with the accuracy shown.

Traditionally, accuracy, when applied to a number, denotes the agreement of that number with the true figure, and precision is the close agreement between measurements taken repetitively. These interpretations, although widely used, do not confirm with existing standards concerning calibration, and care should be used to avoid misinterpretation.

The *error of a number* representing a measurement *can be as much as, but never more than, half of the last significant digit.* The error can be either positive or negative. For example, a measurement of 1235 m implies that the true length is between 1234.5 and 1235.5 m (that is, 1235 + 0.5 and 1235 - 0.5). Other examples:

Written Number	*Implied Limits of the True Value*
203 kg	202.5–203.5 kg
6.312 g	6.3115–6.3125 g
7 ft	6.5–7.5 ft
9.052×10^7 Bq	9.0515×10^7–9.0525×10^7 Bq
63.200 N	63.1995–63.2005 N
52 000 J	51 500–52 500 J (for this number, see Section 1.2.1.5)

If the accuracy or precision of a number is given or implied any other way, it supersedes the "automatic" rule. For example, 236 ± 0.2 m carries its own accuracy guarantee. Taking a 100 Ω resistor, the resistance limits will not be 50 and 150 Ω but the value determined by its color markings or construction.

1.2.1.5 Significant Digits

The accuracy of a number depends on the number of the significant digits: 134 has three significant digits, 2.706 has four. The simple digit count can be used in every case, except when zeros are used to *locate* the number to give its correct positional value (decimal place).

In the measured number

$$213\ 400$$

result of measurement   "positioning" zeros

the "positioning" zeros do *not* count as significant digits. Accordingly, the number above has four significant digits. Since the last significant digit is the "4," one unit of it is 100, half of it is 50, and the limits of the true value are 213 350 and 213 450.

On the other hand, 52 003 has five significant digits; the zeros are true numbers in this case. 3000.0 has five significant digits as well. The decimal point and the last zero are obviously not mandated to give the "3" its positional value; it is the result of the measurement, which, incidentally, produced a number of measured zeros.

In the number

$$0.004\ 23$$

positioning zeros   measured value

the first zeros are exclusively for the positioning of the measured values; accordingly, they do not count in the total. The number above has three significant digits. But in the number 0.012 300 there are five significant digits, because the last two zeros are not there to determine the positional value of the preceding digits the way the zeros on the left do.

For a quick check, find the number of significant digits of the following numbers.

(a)   5340
(b)   3.0150
(c)   0.003 10
(d)   4000.00
(e)   15 000

(a) three;   (b) five;   (c) three;   (d) six;   (e) two

The number of significant digits will not in itself define the accuracy. The implied maximum error of the number 11 m is ±0.5 m, which is 4.55% of the measured 11 m. The implied maximum error of 99 m, the same ±0.5 m, is only 0.51% of 99 m, although both numbers are accurate to two significant digits. Although this must be kept in mind, the number of significant digits does limit the magnitude of the error.

1.2.1.6   Precision

Precision refers to the position of the last significant digit. For example, 175.314 m is more precise than 175.3 m; the first measurement is precise to a thousandth of a metre; the second, to a tenth.

Precision has no relation to accuracy (as represented by the number of significant digits). The value 2.75 has the same precision as 253.81, although the second number is more accurate. Of the two numbers 4326 kg and 3.15 kg, the first is *more accurate* and the second is *more precise*.

For practice, find the more accurate and the more precise numbers of the following given pairs.

(a)   1. 3470 cm   (b)   1. 0.0395 g
      2. 500.0 cm   2. 0.0088 g

(c)   1. 0.293 4   (d)   1.   35 480
      2. 1.020   2. 321 000

(a)   2 is more precise and more accurate
(b)   1 is more accurate; same precision
(c)   same accuracy; 1 is more precise
(d)   1 is more precise and more accurate

1.2.1.7   Manipulation of Numerical Data

When performing mathematical manipulations, great care should be taken that:

1.   No accuracy is lost because of the manipulations.
2.   No spurious accuracy is presumed.

If numbers representing measurements are manipulated, care should be taken to observe these two rules, of which the second is violated more often. Electronic calculators store, manipulate, and show numbers to the limit of their capacity, which is usually in excess of what is justified. For instance, one side of a 2.10-m² rectangle is 1.3 m. How long is the other side? The calculator shows that 2.1/1.3 = 1.615 384 615. It does not mean that we know the size of the side to a billionth of a metre. The original area could be anywhere between 2.095 and 2.105 m² and the given side can be between 1.25 and 1.35 m. Using the extreme values, the other side can be anywhere between 1.552 and 1.684 m. The correct answer is thus 1.6 m, since this is the *best* although *not a perfect* approximation of the possible true size of the side.

For the best retention and proper expression of accuracy, the following rules should be followed.

Addition and Subtraction. When adding and subtracting measured data, the result must be of the same *precision* as the *least precise* of the values used in the calculation. Take as an example a distance measured between two points in three sections:

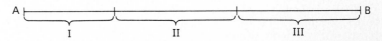

The first person, measuring section I, gets 365.3 m. A second person, measuring section II, gets 415 m. A third person, measuring section III, gets 395.07 m. The total distance is

$$
\begin{array}{r}
365.3 \\
415 \\
+ \quad 395.07 \\
\hline
1175 \text{ m}
\end{array}
$$

which *cannot* be written as 1175.37 m. Since the measurement of the second person carried with it a ±0.5 m implied error, writing down the number as 1175.37 or 1175.4 will be misleading, since it would imply a precision to ±0.005 or ±0.05 m, respectively, which is evidently not the case.

Similarly, adding the measured numbers 567 100, 18 425, and 63, the result will be 585 600, since the last significant digit of the number 567 100 is the "1," determining the *precision* of this number and that of the answer as well.

Doing the calculation, all available digits can be carried, and the result *rounded* to the required number of significant digits. There are rules for the minimum precision to be carried during calculations, but with electronic calculators there seems to be no reason for carrying out intermediate transactions or roundings. The calculator has high speed and great patience.

Carry out the following calculations.

(a)   510.3 kg + 83.45 kg + 100.0 kg
(b)   0.0492 g + 1.032 95 g – 0.005 5 g
(c)   6300 m – 750 m – 153 m

(a) 693.8;   (b) 1.0767;   (c) 5400 m

Note: Rounding numbers is a problem if "5" is the digit to be rounded. The rule says that it should be rounded up or down as necessary to get an *even* number, but computers routinely round 5 up (that is, they add 5 to the first truncated position, then truncate). To be consistent with the machines, in this book 5 will always be rounded *up*.

Multiplication and Division. The rule is: The number of significant digits of the least accurate number determines the accuracy (number of significant digits) of the result.

For example, the sides of a brick are measured as 3.794 cm, 11.26 cm, and 35.4 cm. Its volume is the product of the sides: 1510 cm^3 or 1.51 dm^3 (*not* 1512 or 1512.3 cm^3).

Try these examples (all measured numbers).

$$\text{(a)} \quad \frac{0.0974 \times 1.36}{0.0070} \qquad \text{(b)} \quad \frac{1.003 \times 0.975}{52\,575} \qquad \text{(c)} \quad \frac{1}{0.963^2}$$

(a) 19; (b) 1.86×10^{-5}; (c) 1

1.2.2. SI Units of Measurement

The unit of a physical variable is an arbitrarily standardized quantity. In the SI system there are base units, supplementary units, and derived units. Units can be compound, comprising the names of two or more units and powers (for example m/s, m^2). The unit has a *name* and a *symbol*. For example, "metre" is the name, "m" is the symbol.

A few basic rules about writing names and symbols:

1. Do not mix names and symbols; use Nm or newton metre, not N metre.

2. In text material, a symbol should not be used to start a sentence.

3. Names always start with a lowercase letter, even when derived from a proper name (for example: newton, pascal).

4. Symbols must be printed upright (no italics).

5. The first letter of a symbol is uppercase if it is derived from a proper name (N or Pa, but m).

6. Symbols remain unaltered in the plural (2.5 m, not 2.5 m's or ms).

7. Symbols must have no periods (except at the end of a sentence).

8. There must be a space between the numerical value and the name or symbol: 32 m, 5×10^6 Bq—with one exception, 75°C.

9. Always use *symbols with* numbers, *names without* numbers: 36 kg; potatoes are sold by the kilogram.

Table 1.1   SI Prefixes

Multiplying Factor	Prefix	Symbol
1 000 000 000 000 000 000 = 10^{18}	exa	E
1 000 000 000 000 000 = 10^{15}	peta	P
1 000 000 000 000 = 10^{12}	tera	T
1 000 000 000 = 10^{9}	giga	G
1 000 000 = 10^{6}	mega	M
1 000 = 10^{3}	kilo	k
100 = 10^{2}	hecto	h
10 = 10^{1}	deca	da
0.1 = 10^{-1}	deci	d
0.01 = 10^{-2}	centi	c
0.001 = 10^{-3}	milli	m
0.000 001 = 10^{-6}	micro	μ
0.000 000 001 = 10^{-9}	nano	n
0.000 000 000 001 = 10^{-12}	pico	p
0.000 000 000 000 001 = 10^{-15}	femto	f
0.000 000 000 000 000 001 = 10^{-18}	atto	a

1.2.2.1   Multiples of SI Units

Instead of using zeros, scientific notation, or engineering notation, SI unit multiples or submultiples are given with prefixes. The
prefix precedes the unit without a space between them. Use the prefix names with unit names, the prefix symbols with unit symbols.
The SI prefixes are shown in Table 1.1.

The following are guidelines for using prefixes:

1. Compound prefixes should not be used: nm, not mμm;
    Mg, not kkg.

2. Prefixes cannot be mixed: 8.625 m, *not* 8 m 62 cm 6 mm.

3. The prefixed multiple is usually chosen so that the numeri-
    cal value will be between 0.1 and 1000. This rule should be
    followed reasonably. For example, in a calculation compar-
    ing masses, 15 200 kg can be carried; in a text discussing
    distance in kilometres, 4520 km rather than 5.42 Mm is
    given. Note that mV/mm equals V/m. Prefixes should be
    simplified as much as possible.

1.2.2.2   Base Units

There are seven base units in the SI system (Table 1.2). There
are two supplementary units: for the plane angle, the radian (rad),
and for the solid angle, the steradian (sr), neither of which is discussed in this text.

Following are definitions of the base units:

Metre: the length equal to 1 650 763.73 wavelengths in vacuum
of the radiation corresponding to the transition between the
levels $2p_{10}$ and $5d_5$ of the krypton-86 atom.

Table 1.2   Base Units in the SI System

Quantity	Name	Symbol
Length	metre	m
Mass	kilogram	kg
Time	second	s
Electric current	ampere	A
Thermodynamic temperature	kelvin	K
Amount of substance	mole	mol
Luminous intensity	candela	cd

Kilogram: the mass of the international prototype (étalon) kept in Sèvres (France).

Second: the duration of 9 192 631 770 periods of the radiation corresponding to the transition between the two hyperfine levels of the ground state of the cesium-133 atom.

Ampere: the constant electric current which, if maintained in two straight infinitely long parallel conductors of negligible cross section exactly 1 m apart in a vacuum, will produce between these conductors a force equal to 2×10^{-7} N for each metre of length.

Kelvin: the fraction 1/273.16 of the thermodynamic temperature of the triple point of water.

(Definitions of the mole and the candela are given in part I of the Appendix. These units are not used in this text or in electrical measurements.)

Of the basic units defined above, only the kilogram is to be copied from the original prototype. The other units are *independently reproducible*. This greatly facilitates the maintenance of high standards.

Table 1.3 lists the units that were taken over and incorporated into the SI system.

Table 1.3   Units Incorporated into the SI System

Quantity	Unit	Abbreviation	Equivalent To:
Time	minute	min	60 s
	hour	h	3600 s
	day	d	86 400 s
	year	a	
Plane angle	degree	°	$(\pi/180)$ rad
	minute	′	$(\pi/10\ 800)$ rad
	second	″	$(\pi/648\ 000)$ rad
Volume	litre	l or ℓ	1 dm^3
Temperature	degree Celsius	°C	0°C = 273.15 K (*not* °K!); for temperature intervals 1°C = 1 K

Table 1.4 Common Derived Units in the SI System

Quantity	SI Unit		Expressed in Terms of Other SI Units	Expressed in Terms of Base and Supplementary Units
	Name	Symbol		
Frequency	hertz	Hz	s^{-1}	s^{-1}
Force	newton	N	$m \cdot kg/s^2$	$m \cdot kg \cdot s^{-2}$
Pressure, stress	pascal	Pa	N/m^2	$m^{-1} \cdot kg \cdot s^{-2}$
Energy, work, quantity of heat	joule	J	$N \cdot m$	$m^2 \cdot kg \cdot s^{-2}$
Power, radiant flux	watt	W	J/s	$m^2 \cdot kg \cdot s^{-3}$
Quantity of electricity, electric charge	coulomb	C	$s \cdot A$	$s \cdot A$
Electric potential, potential difference, electromotive force	volt	V	W/A	$m^2 \cdot kg \cdot s^{-3} \cdot A^{-1}$
Electric capacitance	farad	F	C/V	$m^{-2} \cdot kg^{-1} \cdot s^4 \cdot A^2$
Electric resistance	ohm	Ω	V/A	$m^2 \cdot kg \cdot s^{-3} \cdot A^{-2}$
Electric conductance	siemens	S	A/V	$m^{-2} \cdot kg^{-1} \cdot s^3 \cdot A^2$
Magnetic flux	weber	Wb	$V \cdot s$	$m^2 \cdot kg \cdot s^{-2} \cdot A^{-1}$
Magnetic flux density	tesla	T	Wb/m^2	$kg \cdot s^{-2} \cdot A^{-1}$
Inductance	henry	H	Wb/A	$m^2 \cdot kg \cdot s^{-2} \cdot A^{-2}$
Luminous flux	lumen	lm	$cd \cdot sr$	$cd \cdot sr$
Illuminance	lux	lx	lm/m^2	$m^{-2} \cdot cd \cdot sr$
Activity of radionuclides	becquerel	Bq	s^{-1}	s^{-1}
Absorbed dose of ionizing radiation	gray	Gy	J/kg	$m^2 \cdot s^{-2}$

1.2.2.3 Derived Units

The more common derived units, given in Table 1.4, have their own special names and symbols. Other derived units, without special names, are listed in part II of the Appendix.

There are other units not related to SI which are, for the time being, permitted to be used with the SI system (see part III of the Appendix). Some old metric units, however, should *not* be used with the SI system (see part IV of the Appendix).

1.3 ACCURACY OF MEASUREMENTS

Measurement is comparison. Generally, an unknown quantity is compared to a known one. The known quantity can be a special scale or gauge or a general-purpose measuring device. If it is general purpose, it is marked with SI units or with surviving traditional ones.

No measured value can be perfectly accurate. The reasons for measurement errors are:

1. The measuring tool or instrument is not perfectly accurate.
2. The setup for comparison (measurement) is inadequate.
3. The reading of the scale (or equivalent) is limited.

1.3.1 Instrument Accuracy

Measuring instruments are graded according to their accuracy. *Primary standards* are measuring instruments established according to the principles described in Section 1.2.2.2 (except for the mass, which has the one and only primary standard in Sèvres).

Secondary standards are measuring instruments compared directly to primary standards, with setup and reading errors kept below a prescribed minimum.

Tertiary standards are measuring instruments compared to secondary standards; and so on. For all these steps rigid rules of comparison apply and, consequently, the accuracy of the resulting measuring instrument is guaranteed as given. Any measuring tool or instrument so obtained has what is called a *traceable accuracy*, or briefly it is stated that "the instrument is traceable."

> *Note:* In general usage *accuracy* and *error* are used with some overlap. Accuracy is given by naming the magnitude of the error. If a rod with a 1 metre marking is 98 cm long, its accuracy is *not* given as 98% accurate (or 98 cm accurate), but it is noted that the error is 2 cm or 2%. Sometimes, it is said to be 2 cm or 2% accurate. In this text, all accuracy references, whether noted separately or not, are given in *error* units.

To be able to verify the accuracy of a measurement, the accuracy (that is, the maximum possible error) of the measuring instrument must be known—either as established by the user or guaranteed by the supplier.

1.3.2   Accuracy in Setting Up

To avoid errors by faulty setup, care and prudence must be used (such as avoiding badly aligned measuring tape, leaking pressure connections, and faulty electrical connections). But since the number of errors depends on the measuring situation and the carelessness of the measurer, it is difficult to give more specific rules than the following: *Be careful! Be neat! Take your time! Discard questionable results!*

The accuracy of the reading of analog scales depends on the resolution. Resolution is the number of digits directly readable plus an *accurate* estimate of the last digit (rarely, digits), if possible.

1.3.3.   Reading Resolution

Reading resolution using *analog scales* depends on the number of divisions and subdivisions—the thickness and neatness of the marks. If it is a measuring instrument with a pointer, it also depends on the shape of the pointer and the method of avoiding parallax reading error.

Figure 1.1 shows a measuring scale lined up with a piece to be measured. The length can be read as "3.62 units." It would be far-fetched to try to estimate the hundredths. With another measuring

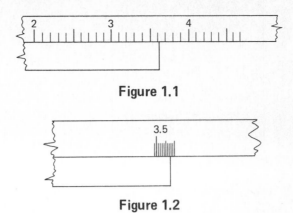

Figure 1.1

Figure 1.2

tool (Figure 1.2) the reading can be made as "3.650 units," although the last digit may not be entirely accurate. A vernier would permit a still more accurate reading on the same scale. (Since the measurements in this book do not regularly use vernier-type readings, the vernier will not be discussed here.)

Scales sometimes cannot be placed right beside the variable or pointer indicating the quantity. The rule is that the reading must be made *perpendicular to the scale*. A reading at an angle introduces an error, as shown in Figure 1.3. Some instrument pointers are turned up at 90° to facilitate alignment. Good instruments have a mirror at the scale to permit the alignment of the pointer with its image, which guarantees no parallax error (Figure 1.4). Gauge manufacturers usually combine a high-grade, accurate mechanism with a well-constructed scale and pointer. The lack of these is usually a statement about the general accuracy of the measuring instrument.

1.3.4. Errors in Instruments

Simple measuring tools and instruments, such as a measuring tape, glass thermometer, or U-tube manometer, carry their own accuracy—which is as good or bad as they were manufactured with. But unless damaged, the instrument can be relied upon in regular use.

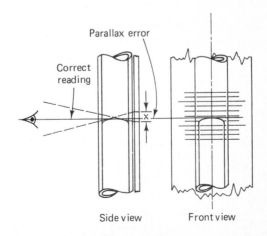

Figure 1.3

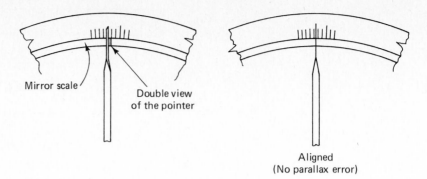

Figure 1.4

More complicated instruments, which may consist of a measuring element, linkages, readout, and adjustments (instruments such as D'Arsonval meters, pressure gauges, transmitters) have many ways to go wrong and have varied sources of error. These errors can be grouped according to their nature and origin.

Unlike their simple counterparts, these instruments can lose their accuracy in use (or abuse), sometimes gradually, without any sign of damage. It is sometimes necessary to *adjust* them for best performance.

If it is necessary to establish the accuracy of such an instrument at any point, it can be *calibrated.* Briefly, we shall recognize the *types of errors*, learn how to minimize them by *adjustment*, and learn how to measure and document those errors by *calibration.*

1.4 ADJUSTMENT AND CALIBRATION

1.4.1 Types of Errors

Errors can be caused by carelessness, misreading, stupidity (more politely: lack of proper knowledge for the particular application), or a misused or damaged instrument. The name of this group of errors is *gross error.* Avoid making such errors.

Of the other errors, called "regular" errors, some are caused by improper adjustment or by specific behavior patterns of the instrument. This type of error is always in the same direction and of comparable magnitude. Such errors are called *systematic errors.*

Some errors are caused by the nonrepeatable operation of the instrument, usually stemming from construction problems or just low quality. These errors can occur in either direction in differing ways, and they are called *random errors.*

Adjustment and calibration are performed to improve and document the accuracy of an instrument. Accuracy does not include any reading and/or setup (gross) errors, which may add to the instrument error to constitute the total reading error. Since the reading of an instrument and arranging the measurement are highly individual, the error cannot be documented. Let's just say that some people are better at measurement than others. Try to be good at measurement.

If one is interested in *adjusting* an instrument, the systematic errors are grouped to facilitate the adjustment into three groups. These are:

1. Linearity error
2. Span error (gain error)
3. Zero error

Calibration groups all errors for the convenience of quality determination and reading correction as:

1. Reference accuracy
2. Hysteresis
3. Repeatability

1.4.2   Instrument Adjustments

The rules of adjustment and calibration are valid for all instruments regardless of their purpose or complexity. The input and output units can be identical or different (as, for example, a signal converter with an input range of 20 to 100 kPa and an output range of 4 to 20 mA).

To check the alignment of, for example, a pressure gauge with a range of 0 to 50 kPa, the input is connected to an accurate pressure source. The input is varied between 0 and 50 in roughly equal steps, and the output is noted at each point.

> *Note:* To keep the discussion simple, the accuracy of the input is accepted as "very accurate" without any numerical reference. If this is not the case, the response has to be established from a calibration-type series of measurement runs.

The input–output relationship is drawn on a Cartesian graph. For example, the measurements (in kPa) are:

Accurate Input	Measured Output
0.0	6.0
10.0	18.8
20.0	27.6
30.0	35.1
40.0	41.3
50.0	46.8

Figure 1.5 shows these values represented on a Cartesian graph. (The errors are exaggerated to make the point.)

A perfect instrument would show a straight line corner to cor-

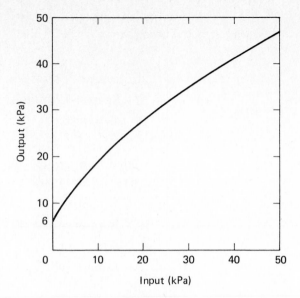

Figure 1.5

ner. As can be seen from Figure 1.5, this instrument has a linearity error (the line is curved), a span error (the average slope of the line is not 45°), and a zero error (the line starts at 6 kPa).

A few more definitions are necessary before we continue. The *lowest value* on the scale is called *zero*. It is 0 in our example, but the pressure gauge could read 100 to 300 kPa, in which case the "zero" value would be 100 kPa.

Span is the extent of units the instrument is able to measure, that is, the difference between the highest and lowest measured values. For the instrument in Figure 1.5, the span is 50 kPa (50 - 0 = 50); for the other instrument mentioned above, the span is 300 - 100 = 200 kPa. *Range* is the designation of the *highest and lowest measured* values; it is always two figures. For the first instrument the range is 0 to 50 kPa; for the second, 100 to 300 kPa. Notice that *span* is always increasing when the *sensitivity* or *gain* is decreasing and vice versa.

Based on this information, the adjuster knows what adjustments are necessary. As a rule, adjust:

First: the linearity
Second: the span
Third: the zero

because span adjustment throws off the zero and linearity adjustment throws off both the span and zero, but not if it is done in the way shown.

1.4.2.1   Linearity Adjustment

Measuring instruments, in contrast to simple measuring tools, convert the measured variable—sometimes more than once—to display the measured quantity in the desired way. For example, in a pressure gauge, the pressure distorts a special sensing tube (a Bourdon tube); the movement of the tube is transferred to an amplifying gear mechanism, the gear mechanism moves the pointer, and the alignment of the pointer with the scale lets one read the quantity.

If everything goes well, the conversion or conversions are linear. In some instruments, when the use of a sensing element with a non-linear output is inevitable, the results can be linearized by a complementary nonlinear component; or in digital instruments or data handling, the variable can be corrected by calculated correction factors.

Good instruments are designed and constructed to be linear. Cheap ones may be nonlinear without any possibility of correction or adjustment. There is, actually, only one category where distortions can occur and adjustments can be made—when mechanical linkages are used for transferring motion. The problem comes from:

1. Converting a rotating motion to a linear one, or vice versa; or

2. Transferring a rotating motion with unequal radii

As an example, look at Figure 1.6. The motion transferred from the wheel to the rod will be dependent on the angle between the rotating arm (r) and the transfer rod (l). With an infinitely long arm, the linear motion will be $r \cos \alpha$, where α is the angle between the rotating arm and the line of the motion. That is, the motion will be maximum and the movement of the transfer rod equal to the movement of the rotating arm when $\alpha = 90°$.

With a limited transfer rod length, the motion

$$\Delta l = r \cos \Delta \alpha + \sqrt{l^2 - r^2 \sin^2 \Delta \alpha}$$

which shows (bypassing the mathematical analysis) that the movement of the rod end will be equal to the movement of the end point of the rod when the rotating arm and the transfer rod are perpendicular.

It shows that the accuracy of such a mechanism is perfect only when a right-angle condition exists, and that the longer the arc of

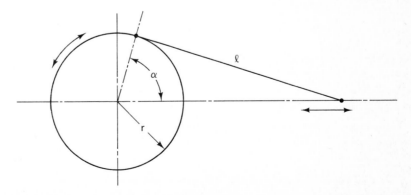

Figure 1.6

movement used, the larger the nonlinearity error becomes. The lesson of the story is that the two arms must be perpendicular at the point where maximum accuracy is required.

As a basic construction principle: *Measuring instruments are constructed in such a way that the accuracy is best at 50% of the measured range.* (This rule is valid for perfectly linear instruments as well, for reasons explained later.)

Figure 1.7 shows the transfer of two angular motions. In the position shown, when both takeup arms are perpendicular to the cross-link (transfer arm), a limited movement of point A will result in an identical movement of point B, and the angular difference of the takeup arms will depend solely on the radius ratio r_1/r_2. Since the r_1/r_2 ratio was established to get the correct span of the instrument, angular movements away from the position shown (right angles) will increase the angular movement of the smaller arm above that of the longer one.

The *rule of adjusting linkages* is:

1.   Adjust the input to 50% (half range).
2.   Adjust *all* moving linkages to be at right angles with each other. (In Figure 1.7, adjustment "L" will do it.)

(There are special linkages to minimize the linearity error, discussed in books about mechanisms and kinematics.)

1.4.2.2   Span Adjustment

On the mechanism shown in Figure 1.7, moving adjustment "S" away from the center of rotation will increase the difference in angu-

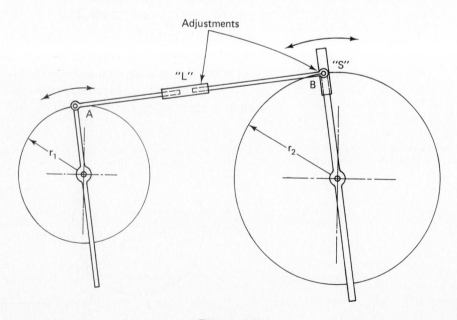

Figure 1.7

lar movements; moving it closer (making r_2 closer to r_1) will decrease the difference.

That is the only specific advice given. To *adjust the span*, two things must be done:

1. Read the manufacturer's instructions about adjustments.
2. Use your brain (because the adjustments are not always crystal clear).

Besides, if you go the wrong way, it will soon be clear.

1.4.2.3 Zero Adjustment

For zero adjustment the same rules apply as for span adjustment. There is a problem, though, if there is a serious noncorrectable linearity error present. If this is the case, the span and zero adjustments can be made in such a way that there will be no error at 0% and 100% reading. The response of the adjusted instrument will resemble that shown in Figure 1.8A and is called *terminal-based linearity*. The disadvantage of this type of adjustment is that the error may be great at the point of intended usage. The response can be adjusted as shown in Figure 1.8B. The name for this is *zero-based linearity*. This adjustment splits the error and maintains the true zero; but to minimize the error to any one side (that is, to divide it equally into positive and negative halves), the adjustments can be made to produce a response such as the one shown in Figure 1.8C. The name for this is *independent linearity*. This type of adjustment will have the smallest value for *maximum* error, but the lack of true zero and maximum reading may be troublesome.

Although the technical specifics for span and zero adjustment depend on the individual instrument handled, there are a few commonsense rules to follow in all cases:

1. Adjust the linearity as well as you can. (This point does not refer to instruments without adjustable linearity or with proven linear characteristics.)
2. Set the input to 50% of the span and adjust the zero *roughly* to read the corresponding value. (That is, for the instru-

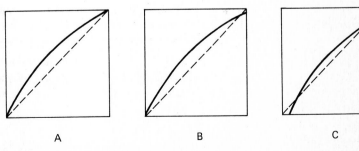

A   B   C

Figure 1.8

ment shown in Figure 1.5, adjust the input and the reading to 25 kPa.)

3. Adjust the input to 25% and then to 75% of its value. Note the output at both settings. If the output difference is the same as the input difference (50% of the span), the span is correctly adjusted. If it is different, adjust the span until a 50% input change results in a 50% output change.

Note: The reason for not adjusting the span using 0% and 100% readings is that because of a possible zero error or non-linearity, there is a "hang-up" at either end. On the other hand, instead of the 25-75 input limits, 20-80 or 10-90% can be used, with a corresponding 60% or 80% output span change. This gives somewhat better accuracy and less convenient span ratio calculations.

4. Repeat step 2, but make it very accurate this time.
5. Check the instrument response between 0% and 100% readings.

With experience—and luck—this is it. In practice, it has to be repeated—and repeated—until the adjustment is as good as the instrument's specifications state.

In most cases, this is the end of the road. If there are errors that are impossible (or not worth) getting rid of, an input-output graph like the one in Figure 1.5 can be completed and kept in the file for reference. If, for example, an official or legal statement of the accuracy of the instrument must be made, after the adjustment a calibration can be performed.

1.4.3   Calibration

Calibration will not *improve or change the operation or accuracy of an instrument.* It only legally demonstrates its accuracy. Adjusting the instrument under calibration is not permitted, and any subsequent adjustment will void the calibration.

Calibration is performed when selling or buying the instrument, when selling or buying products measured by the instrument, or when conforming to legislated accuracy requirements. The accuracy of the instrument is determined according to the established standard, ANSI/ISA S51.1-1979, "Process Instrumentation Terminology."

Little is left to the imagination of the calibrator. It must be performed according to the rules, to ensure that the documents produced will have legal validity.

Calibration of an instrument is quite an involved process, and consists of the following steps:

1. Selection of the master instruments and additional components

2. Preparation of the documents

3. Completion of the measurements

4. Evaluation and data presentation

1.4.3.1   Selection of Master Instruments

As explained in Section 1.3.1, the accuracy of each reading depends on the accuracy of the instrument used for comparison, which we now call the master, and the accuracy of the reading. These should be taken into account when calculating the accuracy of the calibrated instrument. Unfortunately, these calculations would complicate the calibration significantly. So instead of including the accuracy and readability of the master instrument in the calculations, a simple rule is followed. It states that the accuracy and the master *must be one decade better than the intended statement of the accuracy* and that the readability of the master must permit the reading to this accuracy. For example, a thermometer of range 20 to 100°C is calibrated. The measurements are recorded and the accuracies calculated to a resolution of 0.1°C. For calibration, a master must be found which *is accurate and readable to 0.01°C.*

> *Note:* It was not stated that the master must be more accurate and readable than the instrument calibrated, which, incidentally, may not be the case. The comparison of accuracies is between the accuracy of the master and the *stated accuracy in the calibration.* When the accuracy of the master is mentioned, it can be accuracy as read on the instrument, or accuracy calculated using correction curves.

We cannot go into the details of procurement and identification of master instruments, but it must be stated emphatically that without a proper master there is no calibration. (The rules mentioned are valid for calibrating instruments for end use, not for calibrating lower-ranked standards of "secondary masters." To do this, different and more stringent rules apply.)

In addition to the master and calibrated instruments, the calibrating setup consists of interconnecting hardware, signal source, and so on. These must be selected, and the interconnections made, in such a manner that no degradation of the signal can occur at any point. Following are some useful hints regarding the calibration setup.

1. *Pneumatic connections.* Make sure that: *all* connections are leaktight; new or *clean* tubing and fittings are used; connections are made *without* nylon tape and with a minimum of joint compound; *correct torque is used on the fittings*; the tubes are correctly sized and *as short as possible*; the signal

source (regulator) can cover the range with a minimum of hysteresis; the master and calibrated instruments are firmly fixed in their *stated calibrated position* (horizontal, vertical, etc.); and the supply air is clean and dry.

2. *Electrical connections.* Stranded wire is soldered at the end or is fitted with a terminal; solid wire was stripped without a nick; wires are of the proper size, color coded, and neatly arranged; terminals are well spaced and *clean*; special terminal protection is used when the impedance of a component exceeds 1 MΩ; proper shielding and grounding is applied to protect against noise and interference; proper protection of human beings and instruments are carried out; and consideration of the impedance of all the components shows that no unwanted loading effect shows in the loop (this can be carried out easily after study of this book).

3. *Temperature measurements.* The time constant of both the master and calibrated instruments must be known (premeasured), and enough time must be allotted for each measurement to arrive at the accuracy that is required for the reading (that is, 4.6τ for 1% accuracy, 6.9τ for 0.1% accuracy, 9.2τ for 0.01% accuracy).

1.4.3.2   Preparation of the Documents

The emphasis here is on the word *preparation.* All documents must be completed, ready for the insertion of data, *before* the calibration begins. These documents are valid only if nothing is altered or added later. Without prepared documents, there is no calibration.

All measured values are inserted in prepared, numbered spaces. There is no rule about the construction and heading of the prepared data sheets, just reasonable prudence.

Heading.   The heading must contain *all* circumstantial and ambient information that may influence the reading. In case of doubt, include rather than omit data. Specifically:

1. Place, data, and hour of calibration.
2. Names of calibrator and helper.
3. Model number, serial number, and tag number of the calibrated instrument.
4. Sketch of the setup.
5. Model number and serial number (or other unique identification) of the master.
6. Ambient conditions (temperature, pressure, humidity, etc., as needed).
7. Notes, explanations, conditions, qualifications, and disclaimers. Use this "fine print" section to "cover" yourself when in doubt.

No.	Dir.	Input PSIG	Measured PSIG	Difference	Notes
1	Increasing				
2					
3					
4					

A

No.	Dir.	Input		Output		Notes
		°F	%	mA	%	

B

Figure 1.9

A sample heading is shown in Figure 1.9. Other examples in this text will not necessarily include all details of the heading.

Measurements. The number and heading of the columns depend on the variable(s) and the amount of data conversion included.

Figure 1.9A shows a heading when the input and output variables are identical. The inclusion of the "Difference" (error in measured units) column is optional. So would be another column, if included, the "Error %." Figure 1.9B shows an arrangement for different input-output variables. The conversion to percentages is handy, but optional. Inclusion of calculated figures for data conversion and error calculation cuts down the number of intermediary documents, but poses a problem in case of errors. Always leave space for notes.

1.4.3.3 Completion of the Measurements

Three series of measurements (runs) with increasing input (*"up"*) and three series with decreasing input (*"down"*) are the *legal* minimum. If the data points are badly scattered, or no clear trend is discernible, the number of runs *must* be increased (for the sake of simplicity, this text will handle three runs only in the example). The number of input values between the minimum and maximum are not fixed, but should number at least five; it is seldom necessary, though, to have more than 10. The higher number is used when nonlinearity is expected.

To start the readings, *go below* the lowest indicated value (if possible). Then go to the lowest input value, *without reversing*. If you overshoot the mark, go back far enough to safely overcome any hysteresis, and try again. It is nice, but *not absolutely necessary*, to adjust the input to *exactly* the nominal value. For instance, if the intended input value is 4 mA, 3.98 or 4.01 will do just as well for input, since the magnitude of the error will not depend on such small differences. Each subsequent measurement must be carried out in a similar manner.

It takes practice, but above all *patience*, to hit the exact value without reversing. Practice will be acquired. Extra patience is needed at the beginning to make up for the limited practice.

After the last "up" measurement is taken, *go past that value*, and then reverse the direction of readings and continue until the necessary number of runs have been completed.

Corrections. During the measurements, or later when arranging the results, suspicions may arise as to the validity of some of the readings. Gross error, careless readings, faults in the equipment, mishandling, and carelessness can all result in readings that carry unusual errors. If the whole work is suspect, start again (but make it right this time).

If only some readings are suspect, the rule is: Based on qualified judgment, *readings can be omitted*, *readings can be repeated, but no entry to the calibration document can be erased, removed, or made illegible*. A remark must be made in the notes stating the reason for omitting or repeating a reading or readings. Removing data may not imperil the technical validity of the work, but it renders the work legally invalid.

After the last reading is obtained, evaluation may begin.

1.4.3.4 Evaluation and Data Presentation

Evaluation consists of:

1. Converting the data
2. Constructing the deviation table
3. Constructing the deviation plot
4. Determining the accuracies

Converting the Data. Error can be expressed in the units of variable measured or as a percentage. Of these, the percentage is generally used; it has the advantage that it is dimensionless, and it solves the problem of converting one unit into another if the input units are different from the output units. The general formula for conversion is:

$$\text{reading }\% = \frac{\text{reading (measured units)} \times 100\%}{\text{span (measured units)}}$$

Example 1.1

If an instrument with the range 80 to 120 kPa indicates 93.7 kPa, what is the output in percentage?

Using the conversion formula, we have

$$\text{pressure (\% of reading)} = \frac{13.7 \times 100\%}{40} = 34.25\%$$

0% reading is 80 kPa; thus 80 kPa (the elevation of the span) does not take part in the calculation—only the pressure above it, which is 93.7 - 80 = 13.7 kPa. The span is, obviously, 120 - 80 = 40 kPa. What is the error of this instrument if the indicated true pressure on the master instrument is 94.3 kPa?

The error is 94.3 - 93.7 = 0.6 kPa (measured units). Expressed in percentages:

$$E = \frac{0.6 \times 100\%}{40} = 1.5\% \quad \text{(full scale)}$$

(The measurement span is the same.)

Although it is not mandatory to express accuracy and error in percentages, in case of doubt, it can prevent conversion errors.

Other unit conversions can be made using any handy guide or set of conversion tables. (The preferred ones are the *American National Standard on Metric Practice* or *Standard for Metric Practice* or the *Canadian Metric Practice Guide*.)

Constructing the Deviation Table. First, the errors of all the readings must be calculated (measured in units or percentage) and a tabulation made putting the errors of each run in subsequent columns. (The first column is the first series of measurements in the up direction, the second column is the series in the down direction, and so on.)

The sign of the errors must be carried. The error is *positive* if the measured value is *higher* than the reference (accurate) value. (Example 1.2 shows the details.)

In some simple cases this is the end of the representation of the data, but it is always recommended to make a deviation plot.

Constructing the Deviation Plot or Deviation Diagram. The deviation plot shows the errors of the instrument in a graphical way. On this plot, the horizontal axis is the *ideal response* (regardless of what this response looks like), and the *errors* at each point are measured on the vertical axis. To show the difference between a regular input-output diagram and a deviation plot, let us first show all measured points (three sets of measurements) on an input-output diagram (Figure 1.10A). The numerical value of the error is exaggerated in this example; a good instrument would show a response very close to the ideal line. The vertical scale of the deviation plot must be chosen in such a way that the plotted curves are neither too flat nor too exaggerated. In Figure 1.10A the measurements show that the

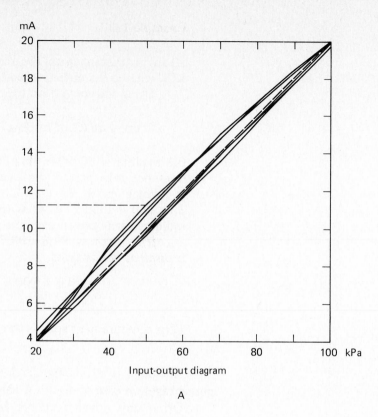

mA

Input-output diagram

A

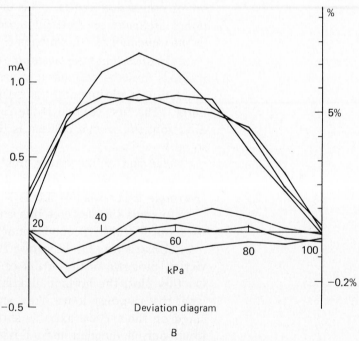

Deviation diagram

B

Figure 1.10

maximum positive error occurs at 50 kPa input, and that its magnitude is 1.2 mA or 7.5%. The maximum negative error is at 30 kPa, and its magnitude is -0.3 mA or 1.9%. Figure 1.10B shows the corresponding deviation plot with a properly chosen vertical scale. The error (vertical scale) can be given in output units, percentage, or (preferably) both. The points are connected with straight-line sections. It is pointless to try to fit curves.

Determining the Accuracies. The measurements displayed in Figure 1.10 permit the determination of the three significant accuracies (or errors) which by general agreement describe the behavior of any instrument. These "accuracies" are:

1. Reference accuracy
2. Hysteresis
3. Repeatability

Reference accuracy is the *maximum error* measured during the calibration runs. It does not matter whether it is positive or negative; it is given *without its sign*. The reference accuracy of the instrument shown in Figure 1.10 is *7.5%*.

In general usage "accuracy" is sometimes quoted without any reference. It always means reference accuracy. Giving the reference accuracy of an instrument based on a proper calibration is a guarantee that at no point will the instrument have an error greater than the one given, regardless of the conditions of the measurement.

Reference accuracy shows directly on the deviation plot (the point farthest from the zero error line). It can be determined from the deviation table; reference accuracy is the entry with the greatest numerical value.

Hysteresis is the *difference* of measured values obtained *at the same input value* between the *up* and *down* directions of the *same run*. Hysteresis is caused by mechanical friction, loose connections, or solid-state devices requiring a minimum forward voltage for conduction. Examination of Figure 1.10B reveals that the greatest distance between points of the same run occurs at 50 kPa, where the distance between the points of the third run is 1.4 kPa or 0.9%. (Watch out! Giving the hysteresis as 0.875% would imply too much resolution!) Hysteresis is given without naming the position of occurrence. It guarantees that at no point of measurement will a greater difference show between measurements taken in the up and down directions.

Repeatability is the *scattering* of the measurements of *different* runs at the *same input point* in the *same direction*. Examination of Figure 1.10B shows that the greater difference exists at 30 kPa (down), at 50 kPa (down), and at 70 kPa (up); at each point the measured values are *0.4 kPa* or *0.25%* apart. Repeatability is given without reference to the point of occurrence. It guarantees that measurements taken *under the same circumstances*, *in the same direction*, will not differ by more than the given figure.

Hysteresis and repeatability can also be determined directly from the deviation table, but great care has to be exercised to spot the right values.

Note: The question arises as to how well the results of as few as three measurement runs can be used as a base for guaranteeing numerical values. According to practice, when *good instruments* are calibrated with *great care*, the results are satisfactory. If the scattering of data were random (Gaussian distribution), mathematical statements could be made accordingly; but only the repeatability may (and not necessarily must) be the result of random scattering. As a result, mathematical analysis of the results is quite pointless.

Example 1.2

The following calibration is performed on an electronic pressure transmitter rated for 0 to 50 kPa input.

    Calibration.   Nov. 13, 1982, 2:30 p.m., Lab. no. 3, ACME Industries

Calibrator: John Doe, CIT, helper Peter Smith

Calibrated instrument: Great Inst. Co. electronic pressure transmitter model no. PT7301, serial no. 1200380, tag no. PT-A214

Calibration setup:

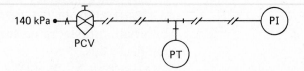

PI: Mercury U-tube manometer, 50 cm length, model no. UP1-3375, serial no. 003984. Scale: 0–70 kPa

PCV: Fisher pressure reducer, model no. 67, tag no. PCV-375, no serial no.

Room temperature: 21°C

Atmospheric pressure: 101.20 kPa

The instrument was calibrated in a vertical position mounted on a pipe. Input was applied through a P3-7 filter. The weathertight cover was on the instrument.

    Readings.

No.	Input kPa	Input %	Output mA	Output %	Notes
1	0.1	0.2	3.65	-2.19	
2	10.1	20.2	6.95	18.44	
3	20.0	40.0	10.25	39.06	
4	30.1		13.55		
5	39.8		16.65		
6	50.0		20.00		
7	50.0		20.50		
8	40.0		17.30		

	Input		Output		
No.	kPa	%	mA	%	Notes
9	30.1		14.03		
10	19.9		10.66		
11	9.8		7.25		
12	0.1		4.13		
13	0.0		3.60		
14	10.0		6.90		
15	20.1		10.23		
16	30.0		12.65		
17	40.0		16.80		
18	50.1	(calculate the rest)	20.13	(calculate the rest)	
19	50.0		20.30		
20	39.9		17.17		
21	29.9		13.85		
22	20.0		10.50		
23	10.0		7.30		
24	0.0		4.00		
25	0.0		3.75		
26	10.0		7.00		
27	19.9		10.12		
28	30.0		13.50		
29	40.1		16.83		
30	50.0		20.10		
31	49.9		20.37		
32	40.0		17.12		
33	30.0		13.85		
34	20.1		10.48		
35	10.0		7.30		
36	0.0		4.00		

Before continuing, calculate and fill in the percent input and output columns of the table. Remember that to get the output percentage, 4 mA has to be subtracted from the measured signal before multiplying it by 100/16.

Next, the deviation table is completed by subtracting the output value (%) from the input value (%).

Input	Error (%)					
(kPa)	1 Up	1 Down	2 Up	2 Down	3 Up	3 Down
0	−2.39	+0.61	−2.50	0.0	−1.56	0.0
10	−1.76	+0.71	−1.88	+0.63	−1.25	+0.63
20	−0.94	+1.83	−1.26	+0.63	−1.55	+0.30
30	−0.51	+2.49	(−5.94)	+1.76	−0.63	+1.56
40	−0.54	+3.13	0.0	+2.51	−0.01	+2.00
50	0.0	+3.13	+0.61	+1.88	+0.63	+2.51

Now the deviation plot can be completed, as shown in Figure 1.11.

The error scale is in output units *and* percentages; compare the percentages with the results of your calculation.

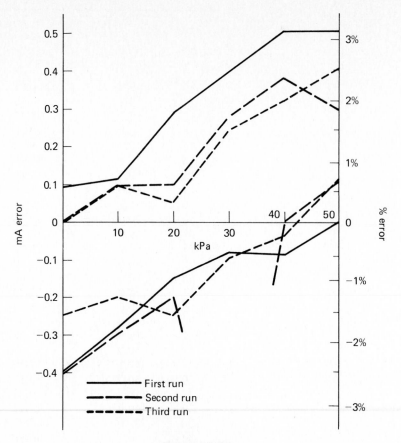

Figure 1.11

From the plot it can be determined that the reference accuracy of the instrument is 3.2%, the hysteresis of the instrument is 3.7%, and the repeatability of the instrument is 1.6%. To be on the safe side legally, all the accuracy numbers were rounded *up*.

Since the location of the maximum error (reference accuracy) is not given, its full value has to be assumed at any point. For that reason, it is sometimes called *full-scale accuracy*.

Example 1.3

The reference accuracy of a pressure gauge is given as 1.5%, and its range is 0 to 60 kPa. As 1.5% of 60 kPa (the span) is $1.5 \times 60/100 = 0.9$ kPa, the measurement *at any point* can be lower or higher by that figure. If this instrument indicates 32.4 kPa, the *true value of the pressure* is then somewhere between $32.4 - 0.9 = 33.5$ kPa and $32.4 + 0.9 = 33.3$ kPa. Note that the *relative accuracy* of the measured pressure decreases as lower and lower pressures are measured. At 60 kPa, the error of the *measurement* is 1.5%. At 30 kPa, the error can be

$$\frac{0.9 \times 100}{30} = 3\%$$

(To find the accuracy at the point of measurement, the 60-kPa span is not used for the calculation, but the actual measured value of 30 kPa.) At 10 kPa, the actual error can be as high as

$$\frac{0.9 \times 100}{10} = 9\%$$

and so on. This is another reason why instruments should be chosen with a range that puts the actual measurement in the upper part of the scale.

1.4.3.5 Use of the Calibration Documents

A look at the deviation plot of Figure 1.11 shows that the instrument is not very accurate, there is considerable hysteresis, and the span adjustment was not done very well. As a result, the following choices of action are possible:

1. Accept the instrument as it is. The accuracy is still satisfactory for the intended use.
2. Send the instrument back for readjustment. In this case the whole calibration has to be repeated.
3. Calculate the averages and attach the average curves to the instruments for reading corrections.

Choices 1 and 2 are self-explanatory. Average curves for correction are illustrated in the following section.

Accuracy Increase with Average Correction.   Assume that during a calibration, at a point where the ideal output reading would be 7.00 V, the three "up" output measured values are 7.15, 7.08, and 7.12 V, and the three "down" values are 7.29, 7.32, and 7.28. This *one set of readings* shows a reference accuracy of 0.32 V, a hysteresis of 0.24 V, and a repeatability of 0.07 V. (Percentage values cannot be calculated because the output span was not given.) Without applying corrections to the reading accuracy at every single reading of the voltmeter, an error of 0.32 V must be assumed.

The *average* (the ordinary mathematical average) of all six readings is 7.21 V. If 0.21 V is added to every 7.00-V measurement, the resulting 7.21 V is only 0.13 V away from the measurement farthest away from this value. The adding (subtracting) of this difference between ideal and average values *improves the reference accuracy of the voltmeter from 0.32 to 0.13 V*, a $2\frac{1}{2}$-fold improvement. The correction for the average could be done either manually or incorporated into any data handling operation as an automatic bias. The average of the three "up" runs is 7.21 V; the average of the three "down" runs is 7.30 V.

If all the values measured at 7.00 V *increasing* are corrected by 0.12 V and all the values at the same voltage measured when the variable is decreasing are corrected by 0.30 V, the *maximum expected error is reduced to 0.04 V* (at the second "up" reading, 7.12 V is assumed when the actual output is 7.08 V). Using the separate up and down corrections, the accuracy is improved eightfold. Application of the hysteresis (up and down) correction is limited, though, to applications when the direction of the measurement is clearly recognizable.

Preparation and Use of the Correction Curves.   Averaging can be done for each input measurement point. The connected average points give the *correction curve.*

If an instrument has good hysteresis (a very small hysteresis error) but otherwise considerable error (because of nonlinearity or bad adjustment), it is sufficient to prepare an average curve to correct all values).

If the error stems mostly from hysteresis (the up and down measured values straddle the zero line at equal distance), only the up and down correction curves can bring improvement. If there is little hysteresis and the error is caused mostly by the scattering of the measured values (a repeatability problem), nothing much can be done to improve the results short of discarding the instrument. That is, the correction curves can be used to correct the influence of *systematic* errors, but not *random* errors.

Example 1.4

Let us continue the preceding discussion and make the correction curves. Since a large part of the error was caused by hysteresis, it is reasonable to make separate up and down correction curves. For the sake of comparison, the overall average is also going to be shown and the improvement in accuracy noted separately.

Figure 1.12 shows the three average curves. The average calculations are not shown. Calculate and check the points on

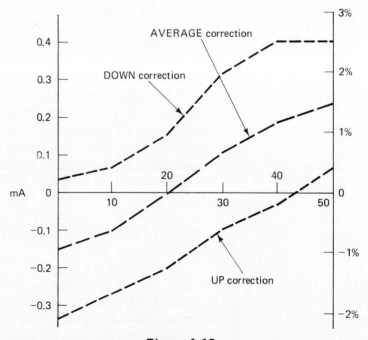

Figure 1.12

the curve. [The second up value at 30 kPa input is left out of all calculations, but the average down value (−1) was included as the sixth entry in the overall average calculation at that point.]

Using the overall average curve to correct the readings, the largest error is 2.0% (0.32 mA). The error is found by comparing the deviation plot with Figure 1.12. The distance between the average curve and the curve of run 1 down at 40 kPa input gives the error. Using the up and down average correction curves, the greatest error of a corrected reading is 0.15 mA (0.9%). Its location is at 20 kPa input, the first down reading.

1.5 PROBLEMS   Solutions to selected problems are given at the end of the text.

1. Name the number of significant digits of each of the following measured (analog) numbers.

   (a) 1230   (b) 4000   (c) 0.0012
   (d) 1.0030   (e) 643 000   (f) 1 250 000.0
   (g) 0.003 740   (h) 10.0001   (i) 0.034 000
   (j) 3 070 040   (k) 0.000   (l) 35 001

2. Find the area of each set of three rectangles.

   (a) 1.250 m × 0.730 m + 12.400 m × 0.017 m +
        1.00 m × 0.6341 m

   (b) 543 000 cm × 7361 cm + 83.615 cm × 526.00 cm +
        45 617 cm × 125.003 cm

3. Convert each of the following measured numbers to a number expressed with the correct prefix.

   (a) 16 000 000 cm   (b) 0.003 10 mm
   (c) 3.15×10^{-7} kg   (d) 4275×10^{9} g
   (e) 6 738 000 nF   (f) 3.15×10^{-4} MJ
   (g) 6720 kPa   (h) 975 kΩ

4. Name the only nonindependently reproducible SI base unit.

5. Find the percentage error of each of the following measurements.

   (a) A length, which is exactly 1.2371 m long, was measured and found to be 1.242 m.

   (b) A mass, 724.70 kg exactly, was measured and found to be 725.0 kg.

6. Name the highest and lowest possible true value of each of the following numbers (the accuracy is quoted for the measured value).

   (a) 16.3 kPa measured with an instrument of 2% reference accuracy

   (b) 2.000 m measured with an instrument of 0.15% guaranteed accuracy

   (c) 0.017 g measured with a scale of 0.25% accuracy

7. Using a set of conversion tables (e.g., from the *American National Standard on Metric Practice* or *Canadian Metric Practice Guide*), make the following conversions.

   (a) 19.36 miles into metres

    (b) 0.725 lb/in^2 (psi) into pascals

    (c) 7 613 100 Btu/hr (international) into watts

    (d) 5.3×10^8 ft lb into watt hours

    (e) 632 lb into kilograms

8. Calculate the highest and lowest possible true value of each of the following measured numbers. The accuracies given are full-scale reference accuracy of the instruments.

    (a) 88.3 kPa measured on a 50-150 kPa 1% accurate instrument

    (b) 135.2°C measured on a 100-150°C 0.5% accurate instrument

    (c) 7.35 mA measured on a 4-20 mA 0.8% accurate instrument

    (d) 17 300 ℓ/min measured on a 0-2500 ℓ/min 1.5% accurate instrument

9. The calibration of a differential pressure transmitter gave the following results. All input and output values are in centimeters of water units.

	Error (%)					
Input	*1 Up*	*1 Down*	*2 Up*	*2 Down*	*3 Up*	*3 Down*
0	-0.040	-0.048	-0.040	-0.070	-0.048	-0.059
2	1.962	1.945	1.964	1.938	1.960	1.942
4	3.972	3.948	3.982	3.950	3.970	3.946
6	5.986	5.950	6.000	5.961	5.980	5.955
8	8.000	7.955	8.013	7.971	7.992	7.968
10	10.011	9.967	10.018	9.987	10.007	9.988
12	12.031	11.983	12.040	12.008	12.021	12.012
14	14.051	14.012	14.059	14.023	14.037	14.037
16	16.080	16.047	16.079	16.052	16.058	16.063
18	18.112	18.087	18.102	18.085	18.084	18.091
20	20.149	20.132	20.126	20.121	20.119	20.130

    (a) Make the deviation table and deviation plot.

    (b) Calculate the accuracies.

    (c) Make correction curve(s) as necessary.

    (d) The following readings were obtained using this instrument. Correct each reading to minimize the error.

        (1) 7.35 cm (up)

        (2) 19.07 cm (down)

        (3) 3.54 cm (down)

10. A temperature indicator with the range 50 to 100°C is calibrated. These are the measured values:

Input	*No.*	*Output*	*No.*	*Output*	*No.*	*Output*	
60	1	60.48	23	60.42	45	60.45	
64	2	64.63	24	60.70	46	64.58	
68	3	68.70	25	68.84	47	68.64	Up
72	4	72.73	26	72.88	48	72.66	
76	5	76.70	27	76.84	49	76.61	

Input	No.	Output	No.	Output	No.	Output	
80	6	80.60	28	80.67	50	80.50	Up
84	7	84.39	29	84.35	51	84.30	
88	8	87.96	30	87.78	52	87.95	
92	9	91.50	31	91.36	53	91.47	
96	10	95.17	32	95.26	54	95.14	
100	11	99.08	33	99.17	55	99.03	
100	12	99.13	34	99.24	56	99.14	Down
96	13	95.21	35	95.34	57	95.26	
92	14	91.38	36	91.48	58	91.41	
88	15	87.65	37	87.67	59	87.64	
84	16	83.99	38	83.88	60	83.92	
80	17	80.23	39	80.08	61	80.17	
76	18	76.35	40	76.25	62	76.35	
72	19	72.42	41	72.37	63	72.45	
68	20	68.44	42	68.46	64	68.49	
64	21	64.42	43	64.52	65	64.47	
60	22	60.34	44	60.52	66	60.42	

(a) Make the deviation table and deviation plot.

(b) Find the accuracies.

(c) Make the (total) average correction curve and correct the following readings.

    (1) $67.2°C$

    (2) $79.1°C$

    (3) $88.0°C$

(d) Make the up and down correction curves and correct the following readings.

    (1) $66.9°C$ (up)

    (2) $93.1°C$ (up)

    (3) $81.6°C$ (down)

    (4) $70.8°C$ (down)

2

The Permanent Magnet Moving Coil or D'Arsonval Movement

The basic instrument used in the following electrical measurement problems will be of the permanent magnet moving coil (PMMC) or D'Arsonval movement type. This movement is well established and widely used in various accuracy grades, alone or in combination with other components. Its use, handling, problems, and behavior are typical. Experience gained with using the PMMC movement can be readily utilized when using other (not necessarily electrical) instruments.

The PMMC movement has an *analog* readout (the numerical value of the measurement is determined by the user, usually by careful interpolation). Instruments with *digital* output are usually easier to read, and the reading error is fixed at half of the last digit, an error value that can be lower or higher than one obtained by a careful analog reading.

Note: The distinction between *analog* and *digital* must be carefully made. The preceding discussion referred only to the readout. If an instrument is labeled "digital," it should have all internal data handling and mathematical operations performed digitally. Analog instruments (those with analog internal operation) may have analog or digital readouts. For the sake of accuracy, the only difference is that with a digital *readout*, the *reading* accuracy is known. Since measurement (as opposed to counting) is the expression of an *analog* quantity into a numerically (therefore *digitally*) defined unit, measuring instruments always have an analog part.

The text and examples will occasionally employ instruments other than the basic PMMC movement (such as volt-ohm meters or electronic voltmeters). Their construction will not be explained, only their use (if it is not self-evident).

Figure 2.1   (Courtesy of Weston Instruments, Inc.)

2.1 CONSTRUCTION OF A PMMC INSTRUMENT

Figure 2.1 is a photograph of a PMMC instrument. A *permanent magnet* creates a magnetic field between the poles. Machining magnets (especially the alloy types) is not easy; in practice, the *pole pieces* are made of soft iron. Between the pole pieces there is a concentric soft iron cylinder. The placement of the iron cylinder in the middle assures a uniform, narrow air gap between the pole pieces and the cylinder. This ensures uniform magnetic field and maximum magnetic field strength in the gap with a given magnet.

The *moving coil* is a winding of copper wire on a square frame (Figure 2.2). The moving coil is positioned concentrically with the iron core and can rotate in the air gap.

There are two spiral hair springs at each end of the coil. They have a dual purpose: to connect the coil to the current source, and to develop a torque if the coil is rotated. To keep friction to a minimum, the pivots are made of heat-treated steel and turn in hard (artificial jewel) bearings (Figure 2.3). The springs are opposed, to minimize the error caused by temperature changes.

The *pointer* is fixed to the coil, usually at a right angle, to show the coil position for the readout. There are *counterweights* mounted on the moving coil for static and *dynamic balancing.*

The outer ends of the spiral springs are fixed in a position to hold the pointer at zero with no current applied. The holding bracket of the outer end of the top spring can be rotated to provide zero adjustment. (The rotation is usually made by providing a "Y" piece which is activated by a pin. This construction assures that the cover, with its adjustment knob, can be taken off without disturbing the

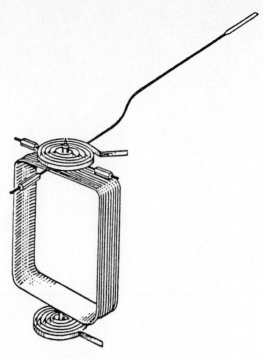

Figure 2.2   (Courtesy of Weston Instruments, Inc.)

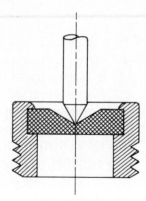

Figure 2.3

movement.) There are stops for the pointer to avoid damage by over-rotating. The frame of the moving coil is made of metal (aluminum). It will provide damping if the coil rotates.

If current goes through the coil, it develops the coil's field. The magnitude of the magnetic field depends on the number of ampere-turns. In practice, it depends on the physical construction of the coil (size, shape, cross-sectional area of the wire, number of turns), and the current applied, in a linear way. (That is, twice the current, twice the magnetic field strength.)

As Figure 2.4 shows, the polarity of the coil is arranged in such a way that the magnetic field developed by the coil causes the coil to rotate clockwise. In the figure a magnet is shown which creates the same *magnetic field* as the coil. The interaction of the two magnetic fields results in a *torque* which tries to rotate the moving coil. As the coil rotates, the deflection of the spiral springs will develop a

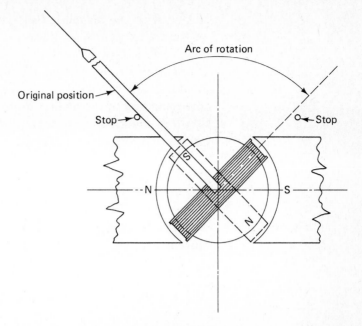

Figure 2.4

countertorque (spiral springs have linear characteristics; that is, the torque linearity increases with the rotation). The moving coil with the attached pointer will rotate until the torque of the springs equals the torque developed by the magnetic field. Ultimately, for a given meter construction, the deflection depends only, and linearly, on the amount of current going through the coil.

Major variations in the construction of PMMC meters are:

1. Instead of pivots and spiral springs, the coil is held by flat strips of spring material ("taut band suspension") tensioned by a spring. This construction eliminates mechanical friction. A disadvantage of this construction is increased size (the band has to be long enough for proper sensitivity) and susceptibility to damage when not handled with care.

2. The magnet is in the core (in place of the iron cylinder) and a soft iron yoke is present around it to assure a uniform magnetic field. For high-efficiency alloy magnets, a considerable savings on space (but not necessarily on purchase price) can be achieved with this construction.

3. Cheap instruments have pin pivots, wider air gaps, and static balancing only. On the other hand, top-quality instruments may have temperature compensation (to hold the meter resistance constant if the temperature varies).

2.2 PMMC METER PARAMETERS

Since current is the only variable that will cause the coil to rotate, the first performance indicator for a PMMC movement is the current needed for full-scale deflection. "Full scale" here refers to the useful range of the coil rotation. As Figure 2.4 shows, when the coil is at a right angle to the centerline of the poles (the equivalent magnet is aligned with the magnetic field), the magnetic force will not result in a rotating torque. The torque is reasonably proportional to the current if the coil (and thus the pointer) movement is limited to about the central 90° movement. Practical scales cover about 90 to 100° pointer rotation. (Smaller arcs give readability problems without any corresponding improvement in linearity.) So when "full-scale deflection" is mentioned, it means that the pointer rotates from the (arbitrarily selected) zero to the 100% points.

The smaller the current needed to deflect the pointer fully, the greater is the *sensitivity* of the meter. *Therefore, the meter sensitivity is expressed as that current* (I_m) *which will deflect the coil by 100% of its selected movement.*

The following variables determine the sensitivity of the PMMC movement:

1. Field strength of the permanent magnet
2. Air gap size
3. Spring constant
4. Number of windings on the coil

The strength of the magnet is a matter of size and cost (for some alloys). Air gap size competes with the number of windings necessary.

The spring constant can be varied between limits and, ultimately, decides the full-scale deflection when the torque created by the magnetic field interaction equals the torque created by the deflection of the springs. Spring size cannot be made very small because (1) it will not carry the current to the coil, and (2) there would be manufacturing and assembly difficulties. A PMMC movement in practical use varies between 10 μA and 10 mA full-scale-deflection (f.s.d.) sensitivity, with the majority of instruments manufactured with sensitivities between 50 μA and 1 mA.

The number of windings determines not only the sensitivity of the meter (and the size of the coil), but also the *resistance* of the instrument. There are installations where the meter resistance is immaterial, and other installations where a higher resistance is advantageous and where it is disadvantageous (as the following examples will show). It is understandable that PMMC instruments are manufactured with different movement resistances. Nevertheless, the *meter resistance* is the second important parameter of the instrument.

A meter with a *higher* sensitivity has a *smaller* f.s.d. current. For practical purposes, the sensitivity is expressed as the inverse of the $I_{f.s.d.}$. From Ohm's law,

$$\frac{1}{I} = \frac{R}{V}$$

and the inverse of the f.s.d. current can be expressed in the unit ohms per volt. It is given as so many ohms per *one* volt.

Example 2.1

A movement has a sensitivity of 2 mA f.s.d. Since

$$\frac{1}{I_{\text{f.s.d.}}} = \frac{R}{1\text{ V}}$$

the meter sensitivity given in this way is

$$\frac{1}{0.002} = 500 \; \frac{R}{1\text{ V}}$$

that is, 500 Ω/V.

Example 2.2

$$I_{\text{f.s.d.}} = 150 \; \mu\text{A}$$

$$\text{sensitivity} = \frac{1}{1.5 \times 10^{-4}} = 6667 \; \Omega/\text{V}$$

Example 2.3

For an electronic meter at a selected scale,

$$I_{\text{f.s.d.}} = 5 \; \mu\text{A}$$

$$\text{sensitivity} = \frac{1}{5 \times 10^{-6}} = 2 \times 10^5 \; \Omega/\text{V}$$

Expressing sensitivity in ohms per volt facilitates some calculations, especially the calculation of ranging resistors; therefore, it is used more often than f.s.d. current. Remember, it is a different expression of the *same* thing.

2.3 SCALES   PMMC instruments are seldom used to measure their "natural" range, that is, the full-scale-deflection current. In most cases, the meter is part of a *measuring circuit.* In place of a scale graduated to show the actual current going through the meter, a *general* scale is supplied which can be used with a proper scale multiplication factor for the value activating the measuring circuit. (Basic movements which are not a permanent part of a measuring instrument have scale multipliers for voltage or current but not for resistance. Movements that are employed as a readout of any given value may have special scales giving the actual measured variable in true units, without the need for a scale factor.) The general scale is usually 0 to 1 or 0 to 100 (sometimes 0 to 10).

Example 2.4

A basic movement is used in a measuring circuit to measure 0 to 25 V; the meter scale is 0 to 1. What is the measured voltage if the pointer indicates 0.374?

Since 25 V is 1 (100% input), the proper ratio is

$$\frac{25\ V}{1} = \frac{x\ V}{0.374}$$

where x is the indication

$$x = \frac{0.374 \times 25}{1} = 9.35\ V$$

(This example is simple enough to be solved by inspection, but remember how to use the ratio if more complicated situations arise.)

Example 2.5

A multirange voltmeter is on the 50-V range. The readout scale is 0 to 10 units, the reading is 8.61. What is the measured voltage?

$$\frac{50}{10} = \frac{x}{8.61}$$

$$x = \frac{8.61 \times 50}{10} = 43.1\ V$$

Example 2.6

A 1.5-mA f.s.d. movement is used without any additional components to measure the 1.5 mA. Its scale is 0 to 1. What should the reading be for 0.705-mA input?

$$\frac{1.5}{1} = \frac{0.705}{x}$$

$$x = \frac{0.705 \times 1}{1.5} = 0.470$$

That is, the input will be 0.705 mA when the pointer indicates exactly 0.470.

To facilitate the conversions, a *scale factor* can be supplied by which each reading can be multiplied to obtain the reading in the desired units. The scale factor is the left side of the original equation, that is, the numerical value of the measuring circuit span divided by the readout scale span. So in Example 2.4 the scale factor is 25 (25/1), in Example 2.5 it is 5 (50/10), in Example 2.6 it is 1.5 (1.5/1).

Calculations are more complicated if *elevated scales* are used. Examples of these are discussed later in the text.

2.4 FINDING THE PMMC MOVEMENT PARAMETERS

Some movements have the f.s.d. current and resistance written or stamped on, others have the f.s.d. current only, and others have no information at all. When an unknown movement is selected for measurement, the first task is to find I_m, the full-scale-deflection current, and R_m, the meter resistance.

2.4.1 Measurement of I_m

2.4.1.1 Using an Ammeter

If there is an ammeter of sufficient accuracy at hand which has a comparable (or suitable) current range, the job is easy. The two ammeters are put in series and the current adjusted until the movement measured shows full deflection, and that current is read on the known meter. Figure 2.5 shows the circuit.

The current is adjusted using a variable resistor. Using an adjustable regulated current source makes the job even simpler—if it can be found at the time of need. The adjustable resistor (or decade box) must have a sufficient value to keep the current below the limit of the (reasonable suspected) minimum I_m of the instrument, and must be adjusted to its *maximum* value before completing the circuit, to prevent meter overloading. The resistance of the ammeter used for comparison must be assumed to be negligible compared to R_r.

For instance, if the construction of the meter suggests a *minimum* I_m of 0.1 mA, and a 6-V dc source is going to be used for the measurement, the *minimum* resistance needed is (from Ohm's law)

$$R = \frac{V}{1} = \frac{6}{0.0001} = 60 \text{ k}\Omega$$

A decade box started at the 59 999 setting or a 100-kΩ potentiometer can be used for current adjustment (preferably multiturn). It is always a good practice to turn the decade box dials to 9 below the intended range adjustment. When reducing the resistance, turning the knob from 1 to 0 will burn out or overload the instrument(s) if the lower-range dial is not set to a high value.

2.4.1.2 Using a Voltmeter and a Decade Box

If no ammeter is available but there is a voltmeter of sufficient quality and a decade box or multiturn pot with a dial, the circuit in

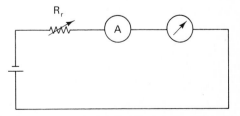

Figure 2.5

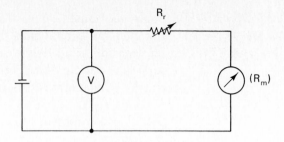

Figure 2.6

Figure 2.6 can be used. R_r should be adjusted (*reduced*) until the meter shows full-scale deflection, then R_r read. I_m then can be calculated from Ohm's law, but the meter resistance R_m must be added to R_r before making the calculation. R_m can be measured first, or the solution of the equation delayed until R_m is known. Since meter resistances are usually between 50 and 500 Ω, the value of R_r is usually much higher than R_m.

2.4.2   Measurement of R_m

2.4.2.1   *Using a Voltmeter and a Ammeter*

The measurement is simplest when a sufficiently accurate voltmeter and ammeter are available. In this case, as shown in Figure 2.5, the measuring circuit can be extended by a voltmeter in parallel with the movement measured (Figure 2.7).

If the voltmeter is an electronic voltmeter with an input impedance of 100 kΩ or higher, the circuit can be set up as shown (with the voltmeter circuit switch closed), the current and voltage read when the PMMC meter shows full deflection, and R_m calculated by Ohm's law.

$$R_m = \frac{V}{A}$$

At full-scale deflection, open and close the voltmeter switch. The measured meter needle should not change position. If it does, it shows a *loading effect*. That is, the resistance (input impedance) of the voltmeter used is low enough to draw a noticeable part of the current into its branch. If there is change but it is small (1 to 2%),

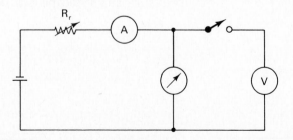

Figure 2.7

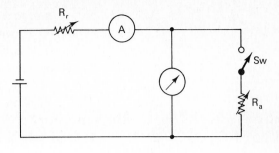

Figure 2.8

determine I_m with an *open* voltmeter switch and readjust the resistor R_r for full-scale deflection when the voltmeter switch is closed.

If the change of indication at the voltmeter switch closing is greater (that is, the voltmeter input impedance is too low for the application), or there is no voltmeter available, the next method should be used.

2.4.2.2   *Using an Ammeter and a Resistor or a Decade Box*

Instead of the voltmeter, a resistor or decade box (R_a) with a switch can be wired parallel to the measured movement (Figure 2.8). First, with the switch (Sw) open, R_r is adjusted for full-scale deflection and the ammeter read, then the switch is closed and R_a adjusted until the meter reads *half scale*. At this point R_r, and then R_a, have to be readjusted until the ammeter shows *exactly* the original reading and the meter is *exactly half* scale. R_a must be read (if it is a decade box or a multiturn potentiometer with a dial) or accurately measured. *The adjusted value of R_a equals the meter resistance*, since the meter deflection was exactly half, so an equal amount of current went through R_a as through the meter. According to Ohm's law, in a parallel connected component (when the voltage drop is identical) equal current goes through equal resistances.

The value of R_a (or the initial setting of the decade box) must be roughly twice that of the assumed resistance of the meter movement.

2.4.2.3   *Using Neither an Ammeter Nor a Voltmeter*

If neither an ammeter nor a voltmeter is available, a circuit such as the one shown in Figure 2.9 can be used for an approximate mea-

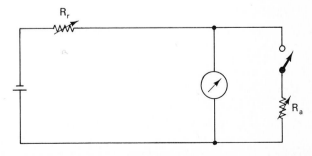

Figure 2.9

surement of the meter resistance. The rules of selection for R_r and R_a are as explained above. First, adjust R_r for full-scale reading. Second, close the switch and adjust R_a until the meter is exactly at half scale; the meter resistance will be somewhere close to the value of R_a. If the nominal source voltage is known, I_m can be then calculated from this voltage and the measured R_m.

The accuracy of this method is limited by the fact that as soon as R_a is switched into the circuit, the total circuit resistance drops, and the current increases. If, for instance, the voltage supply is 1.5 V, the meter movement I_m = 1 mA and R_m = 200 Ω, the movement shows full-scale deflection when R_r is adjusted to 1300 Ω ($R = V/I =$ 1.5/0.001 = 1500 Ω and 1500 - 200 = 1300).

When R_a is switched on and adjusted, the resistance of the complete circuit goes *down* (since R_m and R_a are in parallel). This will *raise* the current through the loop. If the pointer of the instrument is kept at half scale, thus permitting only *half* of the original current through it, the current through R_a must increase, which it can do only if R_a is less than R_m.

This method can be used, but when R_a is measured, the accurate value of R_m must be calculated using the equation

$$R_m = \frac{R_a(2V - R_r I_m)}{I_m(R_r + R_a)}$$

(An explanation of this equation is given in part VII of the Appendix.) In the example given earlier, R_a will be 173.3 Ω when the pointer is at half scale.

I_m also signifies that a certain amount of current is required to deflect the pointer and keep it deflected. That is, a certain amount of energy is required to operate the movement. The circuit measured has to provide that energy. In some cases, this energy can be syphoned off the circuit *without disturbing its operation or the value of the measured variable*.

If the energy drain does disturb the circuit, the effect is called *loading*. For effective use of the PMMC meter, as for *any* measuring device, the user must be sure that there is *no loading* or that the effect of the loading is taken care of when the results are evaluated.

2.5 CARE OF THE PMMC METER

PMMC meters are built according to different specifications and may react in different ways to abuse, but simple reasoning tells us that they need the maximum protection possible from shocks of dropping, rough handling, and blows. In particular, there are a few simple rules to follow:

1. Short the terminals with a piece of wire when in transport or not in use. (The shorted coil develops a counter electromotive force when moving, providing additional damping.)

2. Be sure that the meter is correctly ranged. Overcurrent,

even when it does not ruin the movement, can change its sensitivity and ruin the accuracy.

3. Keep the meter free from water or other moisture, and do not subject it to high temperatures.

4. Always mount the meter securely when in use.

Mounting instructions are sometimes attached to the meter, giving the mounting direction (if any) and mounting compensation (if any). Although good instruments should work in any position (horizontal, tilted, vertical), if an instrument was adjusted, guaranteed, or calibrated in one position, that position may be given by the manufacturer.

Mounting on a steel plate may interfere with the reading. Instruments for panel mounting are sometimes compensated for such mounting, and the instructions will so state (for example: "to be mounted on a 3-mm steel panel"). Such instruments will show an error if mounted in any other way; so will regular (non-precompensated) instruments if mounted on a steel plate without compensation or adjustments. Watch for this.

2.6 OTHER MEASURING DEVICES

In the following discussions PMMC meters were selected to serve as measuring instruments because of their relative simplicity, low price, and limitations. (Better instruments give fewer measuring problems.) It is possible, however, to substitute other voltage-measuring devices if PMMC instruments are not available or if practice with another type of instrument is preferred.

2.6.1 Volt-Ohm Meters

Volt-ohm meters (VOMs) use a PMMC for readout. The movement is directly accessible, bypassing all other internal components, if the meter is switched to its *lowest dc current range.* A 0 to 1 scale has to be selected and used for all measurements. The biggest problem is to resist the temptation to use the existing facilities on the instrument.

The sensitivity of the VOM, which is the same as the sensitivity of its PMMC, is given with the meter, but it should be checked out anyway.

2.6.2 Electronic Voltmeters with Analog Readout

Electronic voltmeters (EVMs or TVMs) have a dc amplifying stage between the input and the readout meter. The readout meter is usually a PMMC movement. In most applications it is possible to select a range (1 mA f.s.d., for example) intended for the measurement. The problem is that the resistance of the unit (which in this case is called the input impedance) is much higher than that of a

PMMC movement, as will be explained in Section 3.4.2). This is especially the case for high-priced instruments.

Many exercises in this book can be completed using an EVM, but the values of the circuit components have to be changed accordingly. Circuits containing high-value resistors must be very neatly constructed, kept clean and dry, and should not be touched during measurements. These meters represent a very low load, which makes it difficult to detect and measure loading effects.

2.6.3   Digital Meters

Digital meters use an analog-to-digital converter to change the analog signal into a digital readout. Digital EVMs combine an amplifier with the analog-to-digital converter. Since solid-state devices need a minimum voltage and current for operation, simple digital panel meters may have an amplifier stage if the measuring range of the meter warrants it. There are meters converting pulses into digital readout, but they cannot be used as substitutes for PMMCs.

Digital meters with *no* or *low* amplification are well suited to replace PMMCs for the work described in this book. Full scale is represented by 9's in all digital positions. Full-scale deflection current and meter resistance should be measured in the same way as for the PMMC meter.

If the resistance of the digital meter is below 1000 Ω, the digital meter can be used as a direct replacement for the PMMC movement. Higher resistances (the same as with EVMs) reduce the loading effect, making laboratory duplication of the book's examples less easy. For digital meters with a very high resistance, a 100-Ω 2-W resistor soldered in series with the meter will convert it so that it has parameters similar to those of a PMMC movement. It is easier to read digital meters because no special care or eyestrain is needed to get the best resolution. With a digital meter, the number of digits determine the resolution (which, by the way, can be the same, better, or worse than that obtainable using an analog meter of comparable quality). Digital meters are more damage resistant and reduce the chance of committing gross errors.

As with analog meters, readings taken at the low end of the range have limited accuracy. Full accuracy can be expected if all digits show numerals. A minimum of four digits is required.

Notice that this comparison covers only resolution and reading accuracy. Whether or not a digital meter is more accurate than an analog meter depends on its design and construction, which is seldom apparent to the user. It must be said that there are more truly unreliable analog meters on the market than digital ones, but it is a grave error to believe that the displayed number on a digital meter represents true accuracy without first checking the meter and its documents.

The rate of change of a digital meter is higher than that of a (damped) analog meter, but it does not show the dynamic conditions (rate of change) of the measurement. An advantage of the digital meter is that rough handling does not diminish its accuracy. If something fails, the meter just goes "blink."

2.6.4 Oscilloscopes

Oscilloscopes can measure periodically changing current, and some models measure dc. Each oscilloscope has frequency limitations. Dc oscilloscopes can measure steady voltage (a battery, for example) and if the frequency is correctly set, peak values of rectified, unfiltered ac. Oscilloscopes have scales to measure voltage directly. The signal is amplified, presenting the same conditions for its application as those of EVMs.

3

DC Voltage Measurements

3.1 MEASUREMENT OF VOLTAGE (POTENTIAL)

The volt is the unit of electrical potential, potential difference, and electromotive force. One volt is defined as the potential difference that produces one watt of power for one ampere current (W/A).

In practice, a distinction must be made between *open-circuit potential* and *voltage drop* across a selected component (or components) in a circuit.

Open-circuit potential shows the presence of a voltage with infinite resistance between the poles, and it can be measured only with an instrument that does not initiate current flow. These types of instruments are called *potentiometers* to indicate their property of measuring without loading.

Voltage can be measured as *open-circuit voltage* (Figure 3.1A) or as a *voltage drop* (Figure 3.1B). If *voltage drop* is measured, the voltmeter is always *in parallel* with the component (or components) across which the voltage drop is to be measured. No circuit alteration is necessary for voltage measurement.

It must be made sure that the range of the voltmeter is properly selected. As mentioned earlier, for accuracy and practicality, best results are obtained if the measured variable is over 50% of the meter range. On the other hand, it must be assured that the meter will not be overloaded by a higher than f.s.d. current going through it. For example, if a voltage source is to be measured which is known to be around 6 V, the meter must be switched to the 10-V range.

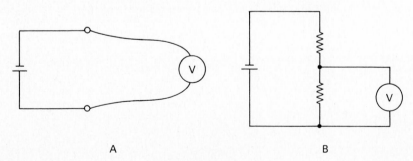

A   B

Figure 3.1

Watch carefully for the difference: If the problem is to prepare a movement for a 6-V *range*, the ranging resistor must be selected to provide just that. But when the problem is given to range an instrument to measure a 6-V voltage drop, it has to be ranged to measure about 0 to 10 V.

3.2   USING THE PMMC MOVEMENT FOR VOLTAGE MEASUREMENTS

Once the I_m and R_m are known, Ohm's law tells what *voltage* will deflect the instrument fully. For instance, for a 0.1-mA f.s.d. 250-Ω resistance meter, the full-scale deflection voltage is $V = I \times R = 0.0001 \times 250 = 0.025$ V or 25 mV.

If the movement is not sensitive enough, that is, a *smaller* f.s.d. voltage is needed, a better instrument (EVM) must be found; the movement on hand cannot do the job. If the measured range required is *larger* than the f.s.d. voltage of the movement, it can be ranged by a ranging resistor, R_r.

The ranging resistor constitutes a *voltage divider*, if it is in series with the movement, leaving just enough voltage drop across the meter for full deflection. Figure 3.2 shows the principle. To calculate R_r, V_m should not necessarily be known. Since the *same current* passes through R_r and the movement (they are in series!), Ohm's law can be written for each part or for the whole circuit. If V is the total voltage to be measured and R is the total resistance of the circuit,

$$I = \frac{V}{R} = \frac{V_m}{R_m}$$

and, by rearranging the equation,

$$R = V \frac{R_m}{V_m}$$

but since $V_m = I_m \times R_m$,

$$R = \frac{V}{I_m}$$

or if the sensitivity is $1/I_m$ ohms per volt, $R = SV$ and R_r is obtained simply as $R - R_m$.

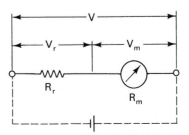

Figure 3.2

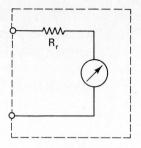

Figure 3.3

Example 3.1

A 1-mA 100-Ω movement is available to make an instrument to measure 0 to 10 V. Find the ranging resistor.

$$R = \frac{V}{I_m} = \frac{10}{0.001} = 10 \text{ k}\Omega$$

$$R_r = R - R_m = 10\ 000 - 100 = 9900\ \Omega$$

or the meter sensitivity S is $1/0.001 = 1000\ \Omega/\text{V}$:

$$R = VS = 10 \times 1000 = 10 \text{ k}\Omega$$

and so on. When this movement–ranging resistor *combination* is referred to later, it will be called the *voltmeter* or voltage-measuring *instrument* (Figure 3.3), in contrast to the *basic movement*.

Example 3.2

A voltage is to be measured which is around 11 V. Make a voltmeter to measure this voltage using a 0.5-mA 200-Ω basic movement.

A full-scale reading of 15 V is selected. (This is an arbitrary decision. For the sake of a simple scale factor, unusual full-scale voltages such as 14, 16, and 18 V are avoided, but a selection of 20 V full scale is acceptable as well.)

$$R_r = \frac{V}{I_m} - R_m = \frac{15}{0.0005} - 200 = 29.8 \text{ k}\Omega$$

Example 3.3

A 2-mA 80-Ω basic movement is to be prepared to measure 0 to 0.75 V. Calculate R_r.

$$R_r = \frac{0.75}{0.002} - 80 = 295\ \Omega$$

Since the actual value of R_r is of no consequence when the voltmeter (not the movement alone!) is used, in calculations using the voltmeter its total resistance (R in the calculations above) is going to be used, notated as R_i.

3.3 MULTIRANGE VOLTMETER FOR DC

If different voltages are to be measured, the usual case when checking an electronic circuit, a multirange voltmeter can be prepared with a switch to select the proper range.

The simplest design is to calculate the value of the ranging resistor for each voltage range, the switch activating the branch with the required resistor. Figure 3.4 shows this design. This circuit has two problems. The first is the probability that all resistors will have odd values. The second is that if the meter resistance were not accurately known, or the movement were exchanged to a similar but not identical one, all resistors would have to be rearranged. To do that, a design such as the one shown in Figure 3.5 can be used.

Although it looks nice on paper, adjustable resistors of the required accuracy and value are not easy to find, and the adjustment requires a master unit with identical ranges. For these reasons, this design is not common, but let's look at an example before proceeding.

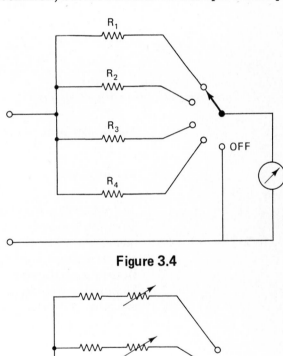

Figure 3.4

Figure 3.5

Example 3.4

Design a multirange voltmeter using a 0.1-mA 300-Ω movement for 0.5-, 1-, 5-, and 10-V ranges. The values are calculated in the same way as for the individual ranging resistors in Section 3.2, and the data are entered in a table:

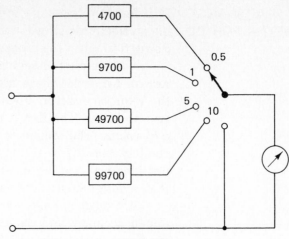

Figure 3.6

	Range (V)			
	0.5	1	5	10
$R(V/I_m)$ (Ω)	5000	10 000	50 000	100 000
$R_r(R - R_m)$ (Ω)	4700	9700	49 700	99 700

The design is shown in Figure 3.6.

Example 3.5

A design which corrects the faults of the design discussed in Example 3.4 is shown in Figure 3.7. R_4 determines the lowest range, $R_3 + R_4$ the next (higher) range, and so on until the highest range, which is determined by the sum of all resistors. Design a multirange voltmeter using the same values as for Example 3.4 (0.5, 1, 5, and 10 V). For now, disregard the adjusting resistor.

	Range (V)			
	0.5	1	5	10
R (Ω)	5000	10 000	50 000	100 000
R_r (total) (Ω)	4700	9700	49 700	99 700
R_4 (Ω)	4700	—	—	—

R_3: $9700 - 4700 = 5000 \ \Omega$ because R_3 $(5000 \ \Omega)$

$+ R_4$ $(4700 \ \Omega)$ equal R_r for 1 V

$= 9700 \ \Omega$

R_2: $49 \ 700 - 9700 = 40 \ 000 \ \Omega$

R_1: $99 \ 700 - 49 \ 700 = 50 \ 000 \ \Omega$

So if the instrument is switched on the 10-V range, the ranging resistance is the sum of all resistors: $50 \ 000 \times 40 \ 000 + 5000 +$

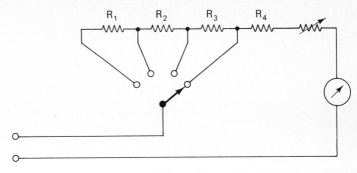

Figure 3.7

4700 = 99 700, as it should be. As can be seen, for quick calculation, take the *difference* of the two adjacent calculated R_r values to find the value of each resistor.

If an adjustable resistor is included, its value combines with the last resistor to make up the lowest range. In Example 3.5, it can be achieved in several ways.

The simplest is to *substitute* R_4 for an adjustable resistor. A 5-kΩ multiturn (wound) adjustable resistor of a 10-kΩ unit can be used (or, anything in between, although multiturn precision resistors are not made for odd values). Simple carbon potentiometers are not easy to adjust or stable enough to be used in this application.

Since only a small part of the resistor is to be used, usually R_4 is made up of a regular fixed-value resistor and a potentiometer. Several combinations can be used, provided that the fixed and the adjustable resistors' resistances add up to the required amount *when the adjustable resistor is set around its center position.*

In Example 3.5, the combination can be: 3000 Ω fixed, 3000 Ω adjustable, *or* 4000 Ω fixed, 1000 Ω adjustable. The important point is that adjusted to the *nominal* resistance, there is adjustment available in both directions on the adjustable resistor.

The only consideration is to include an *available* resistor value, one that is on hand or at least is commercially available. Smaller values for the adjustable resistor make adjustment easier but require more accurate values to start with. Avoid the error of making the sum of the last fixed and adjustable resistors *equal* to the nominal value of the lowest resistance.

The adjustable resistor corrects the *range* error caused by inaccurately measured (or given) movement parameters and the value of the lowest resistor. It should *never* be used for zero adjustment, for which the mechanical zero adjustment is used.

Example 3.6

Calculate the resistors for a multirange voltmeter using a 2-mA 150-Ω movement for measuring 1-, 5-, 10-, 20-, and 50-V ranges.

	Range (V)				
	1	5	10	20	50
R (Ω)	500	2500	5000	10 000	25 000
R_r (Ω)	350	2350	4850	9850	24 850
ΔR (Ω)		2000	2500	5000	15 000

$$R_1 = 15\ 000\ \Omega$$

$$R_2 = 5000\ \Omega$$

$$R_3 = 2500\ \Omega$$

$$R_4 = 2000\ \Omega$$

$$R_5 = 300\ \Omega$$

$$R_{\text{adj}} = 100\ \Omega$$

with other equally good combinations possible for R_5 and R_{adj}.

The design in Figure 3.7 has the resistors on the meter side of the switch, which prevents the use of a shorting contact as was done on the design shown in Figure 3.4. The arrangement shown in Figure 3.7 follows the existing practice of commercial designs.

3.4 VOLTMETER LOADING EFFECT

It was explained that a PMMC movement and any voltmeter using a PMMC movement requires energy for operation. The current consumption of the meter may or may not influence the operation of the circuit and thus the voltages in a given measurement.

When measuring the voltage of a source, the voltage will not change (drop) as a result of the current consumption (loading) of the voltmeter—if the source is *infinitely strong*. There is no such thing, but if the source were strong enough, the voltage drop will be too small to contribute noticeably to the error of the instrument. When measuring a component of a circuit, the voltmeter acts as a parallel resistance with the original component, lowering the resistance, and thereby the voltage across it, causing an erroneous reading (low).

So as to be able to analyze the effects separately, a component in a circuit will first be measured with the assumption that the *source is very strong* and so does not contribute to the loading effect.

3.4.1 Measuring Components in a Circuit

Figure 3.8 shows a circuit with two resistors, R_A and R_B. The task is to measure the voltage drop across resistor R_B.

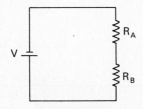

Figure 3.8

This much simplified circuit is adequate to cover all practical in-circuit measurements if it is assumed that R_A *is the total ohmic resistance of all components outside the measured part, and R_B is the total ohmic resistance of all components across which the voltage drop is measured.*

Analysis of a few circuits will show how the circuit and meter parameters influence the error.

In the circuit shown in Figure 3.8, R_A is 1000 Ω, R_B is 1000 Ω, and the supply voltage is 5 V. Measure the voltage across R_B. A 1-mA 100-Ω instrument is available for the measurement.

The selected range of the voltmeter will be 10 V (not 5 V; never try to measure at the top end of the instrument) and the resistance of the voltmeter $(R_r + R_m)$ will be $10/0.001 = 10$ kΩ.

The circuit with the voltmeter attached is shown in Figure 3.9. The voltmeter is attached at points P1 and P2. The *total resistance* between points P1 and P2 is calculated using the equation for parallel resistors:

$$R_x = \frac{1}{\dfrac{1}{R_B} + \dfrac{1}{R_i}} = \frac{1}{\dfrac{1}{1000} + \dfrac{1}{10\,000}} = \frac{10\,000}{11} = 909.1 \ \Omega$$

and the voltage drop between P1 and P2, V_x, will be

$$\frac{5 \ V}{1909.1 \ \Omega} = \frac{V_x}{909.1 \ \Omega} \ , \qquad V_x = \frac{5 \times 909.1}{1909.1} = 2.38 \ V$$

(The left side of the equation states that the loop current is the full voltage divided by the total resistance in the loop; the right side states that it divided the voltage between P1 and P2 by the resistance between P1 and P2.) Since both resistors are 1 kΩ in this example, it is obvious that, without the meter attached to the circuit, resistor R_B dropped exactly 2.5 V.

The addition of the voltmeter distorted the reading to 2.38 V. *This voltage drop was caused by the loading effect of the voltmeter.* The error is

$$E = \frac{100(2.5 - 2.38)}{2.5} = 4.8\%$$

which is more than is permissible.

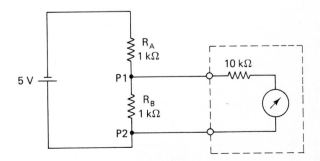

Figure 3.9

Example 3.7

Examine a similar circuit, but one that has both resistors, R_A and R_B, changed to 10 kΩ. The supply voltage is still 5 V, using the same instrument. First, we measure the resistance and voltage between the measured points:

$$R_x = \cfrac{1}{\cfrac{1}{10\ 000} + \cfrac{1}{10\ 000}} = 5000\ \Omega$$

$$V_x = \frac{5 \times 5000}{15\ 000} = 1.67\ V$$

Thus the error is

$$E = \frac{100\,(2.5 - 1.67)}{2.5} = 33.3\%$$

Getting worse. Calculate the error if both R_A and R_B are 100 Ω each. It is better (0.5%).

It can be deduced that for accurate measurement, that is, a small loading effect, the resistance of the voltmeter must be much higher than that of the resistance of the measured part of the circuit. The greater the difference, the less the error.

What would happen if the measurement shown in Figure 3.9 (1000-Ω resistors) were made using a 0.1-mA 100-Ω basic movement. (The 100 Ω here is given only to calculate R_r. Its value for the following calculations is immaterial.) The resistance of the voltmeter will be

$$R_i = \frac{10}{0.0001} = 100\ 000\ \Omega$$

From

$$R_x = \cfrac{1}{\cfrac{1}{1000} + \cfrac{1}{100\ 000}} = 990.1\ \Omega$$

$$V_x = \frac{5 \times 990.1}{1990.1} = 2.49\ V$$

we calculate the error as

$$E = \frac{100\,(2.5 - 2.49)}{2.5} = 0.5\%$$

That is, *having a more sensitive instrument decreased the error,* because it raised the resistance of the voltmeter.

Change R_A to be 100 Ω, but leave R_B at 1000 Ω and bring back the 1-mA 100-Ω meter (Figure 3.10).

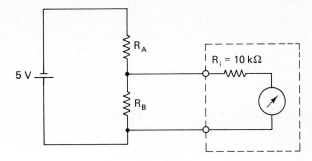

Figure 3.10

$$R_x = \frac{1}{\dfrac{1}{1000} + \dfrac{1}{10\ 000}} = 909.1\ \Omega$$

as before, but

$$V_x = \frac{5 \times 909.1}{1009.1} = 4.50\ \text{V}$$

compared to

$$V = \frac{5 \times 1000}{1100} = 4.55\ \text{V}$$

without the voltmeter attached. The error

$$E = \frac{100(4.55 - 4.50)}{4.55} = 1.1\%$$

The fact that R_A, the resistance outside the measured part, decreased reduced the error. It is logical that if $R_A = 0$, the error will be zero, since the meter now measures the supply voltage (which is in this case, as assumed, very strong).

Increasing R_A, accordingly, will increase the error. Let R_A be 10 000 Ω and R_B 1000 Ω:

$$R_x = \frac{1}{\dfrac{1}{1000} + \dfrac{1}{10\ 000}} = 909.1\ \Omega$$

again, but

$$V_x = \frac{5 \times 909.1}{10\ 909.1} = 0.418\ \text{V}$$

compared to

$$V = \frac{5 \times 1000}{11\ 000} = 0.455\ \text{V}$$

and the error

$$E = \frac{100(4.55 - 0.418)}{4.55} = 8\%$$

It is bad; in addition, we violated the rule that the reading must be about or over 50% of the range. Rearranging the meter to read 1 V full scale, R_i will be 1/0.001 = 1000 Ω. Using this instrument, calculate the error:

$$R_x = \cfrac{1}{\cfrac{1}{1000} + \cfrac{1}{1000}} = 500 \ \Omega$$

$$V_x = \frac{5 \times 500}{10\ 500} = 0.238 \ \text{V}$$

$$E = \frac{100(0.455 - 0.238)}{0.455} = 48\%$$

Now we are between a rock and a hard place. If we add 2% instrument and reading error to the 0.418-V reading on the 10-V range, 2% of 10 V is 0.2 V. The reading (assuming low error) = 0.22 V, and the error

$$E = \frac{100(0.455 - 0.22)}{0.455} = 52\%$$

That is, keeping the reading low to maintain a high voltmeter resistance works only if meter accuracy and readability are excellent. (Some digital instruments may do it.) The purpose of this investigative chapter was to *explain the reason* and *give a feeling* for the loading effect and the possible magnitude of the error. In real life the data given in these examples are not available, and a method must be devised to detect and, if possible, correct for the loading effect.

3.4.1.1 Detection of the Loading Effect

When making a measurement, it is impossible to say what, if any, loading effect is introduced and how accurate the measurement is. If the measurer knows that a very strong source is measured, or the impedance of the instrument is much larger than the resistance of the components across which the measurements are made, the accuracy of the reading can be accepted on faith. In case of doubt, it has to be investigated. Accepting the accuracy of the measured values must be based on solid knowledge of the instruments and circuits involved, *not* on guess, hope, or crossed fingers.

Since the error was introduced by the inclusion of a parallel current path, increasing or decreasing this current, that is, decreasing or increasing the resistance of the attached meter, will change the loading effect and the error.

Although any resistance change in the loop can help to detect the loading effect, to make subsequent calculations easy, the best is to *halve* or *double* the resistance of the voltmeter. This can be done by getting a resistor that has the *same resistance as the voltmeter*, and including it in the circuit in *parallel* or in *series* with the voltmeter.

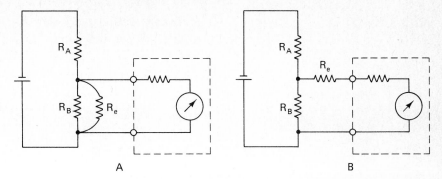

Figure 3.11

The resistance of the home-built voltmeters will be known. For commercial units the specifications will tell the sensitivity (which usually is the same for all ranges), which multiplied with the selected measured range gives the meter resistance for that range.

Figure 3.11A shows the equivalent resistor, R_e, parallel with the voltmeter. Figure 3.11B shows it in series with the voltmeter. Continuing with the example shown in Figure 3.9, R_e will be 10 kΩ. Connect this in *parallel* with the meter and R_B.

The new resistance between P1 and P2 will be

$$R_x = \frac{1}{\dfrac{1}{1000} + \dfrac{1}{10\,000} + \dfrac{1}{10\,000}} = 833.3\ \Omega$$

$$V_x = \frac{5 \times 833.3}{1833.3} = 2.27\ \text{V}$$

$$\left[E = \frac{100(2.5 - 2.27)}{2.5} = 9.2\% \right]$$

The inclusion of the new resistor dropped the voltage and thus increased the error.

Connecting R_e in *series* with the meter, we have

$$R_x = \frac{1}{\dfrac{1}{1000} + \dfrac{1}{20\,000}} = 952.4\ \Omega$$

$$V_x = \frac{5 \times 952.4}{1952.4} = 2.44\ \text{V}$$

$$\left[E = \frac{100(2.5 - 2.44)}{2.5} = 2.44\% \right]$$

As expected, the voltage reading increased and the error decreased.

The observer will not know the numerical value of the change in the error (not knowing the actual undisturbed voltage), but there will be a *0.11-V decrease*, which is a 2.2% change in the full-scale voltmeter reading (100 $\times$ 0.11/5), and in the second case, a *0.6-V increase* (which is a 1.2% change).

In the same setup, if R_A and R_B are 10 kΩ each, the changes will be, with R_e in parallel,

$$R_x = \cfrac{1}{\cfrac{1}{10\,000} + \cfrac{1}{10\,000} + \cfrac{1}{10\,000}} = 3333\ \Omega$$

$$V_x = \frac{5 \times 3333}{13\,333} = 1.25\ \text{V}$$

The voltage *change* will be 1.67 - 1.25 = 0.42 V (a -9.5% change). With R_e in series,

$$R_x = \cfrac{1}{\cfrac{1}{10\,000} + \cfrac{1}{20\,000}} = 6666\ \Omega$$

$$V_x = \frac{5 \times 6666}{16\,666} = 2.0\ \text{V}$$

The voltage change will be 2.0 - 1.67 = 0.33 V (a -6.5% change full scale).

Measuring the voltage change by attaching the equivalent resistor in *parallel* is easy. The existing connections must not be disturbed, and the movement of the meter needle can easily be checked by attaching and removing R_e.

Putting R_e in series requires a careful reading of the voltage, opening the circuit to insert R_e, and another careful voltage reading. The introduction of R_e in series *changed the voltage range of the voltmeter to its double.* The scale factor must be doubled for an accurate reading (that is, the pointer's new position will be somewhat higher than half of the original reading).

The ease of attachment of R_e, the same scale, and the larger voltage change makes the loading-effect test with R_e applied in *parallel* the natural choice. Briefly, to check the loading effect of the voltmeter:

1. Select a resistor with a value equal to the voltmeter.
2. Attach it in parallel with the voltmeter.
3. Watch the meter needle.

If the movement of the meter needle is *hardly visible*, or small enough (-0.5% of the full-scale reading), *the loading effect is small and can be neglected.*

If the change in needle position is between -0.5% and 10% of the full-scale reading, the loading effect *causes a noticeable error for which an adjustment must be made.* If the change in needle position is more than 10%, the sensitivity of the instrument is too low. A more sensitive instrument (EVM) or a potentiometric instrument is required.

The 0.5 and 10% limits given are arbitrary but reasonable with an instrument of 0.1 to 0.5% readability and better than 1% reference accuracy. If the required accuracy is higher or lower, those limits can be lowered or raised, respectively.

3.4.1.2 Correction of the Loading-Effect Error

There are two ways to make corrections for the loading effect. The first uses an adjustable resistor (about double the value of R_i) or a decade box; the second uses an equivalent resistor (R_e), as discussed in Section 3.4.1.1.

The method using the variable resistor is more accurate (most of the error is caused by the inaccuracy of the resistors), but the other method is simpler and still gives satisfactory accuracy.

Variable Resistor Method. The variable resistor can be a decade box or a multiturn potentiometer with a dial. An ordinary variable potentiometer can be used, but its resistance must be measured to the required accuracy (which is usually 1% or better).

> *Note:* Certain methods look good on paper, but their practicality may be limited. Accurate measurement of resistances is not easy unless a good-quality electronic multimeter is at hand. If this is the case, the original measurement could have been made using it, bypassing the problem of a meter with low sensitivity.

The measurement is set up as shown in Figure 3.12. The maximum adjustable value of R_x must be about double that of R_i, the resistance of the instrument ($R_x + R_m$).

To use this method:

1. Set R_x to *zero*, and read the indicated voltage carefully on the meter.

2. Adjust R_x until the meter indication is *exactly half* of the original.

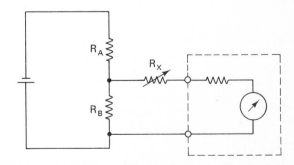

Figure 3.12

3. Calculate the true voltage using the equation

$$V = \frac{V_o \times R_x}{R_i}$$

where V is the true voltage and V_o the original voltage reading of the instrument.

Example 3.8

A 1000-Ω/V voltmeter on the 10-V scale indicates 6.25 V between the two measurement points. A 10-kΩ resistor in parallel with the meter drops the voltage reading to 5.56 V (6.9% full-scale change). To find the true voltage a decade box is put in series with the voltmeter and R_x adjusted until the voltmeter indication is 3.125 V (half of 6.25). At this point the resistance of the decade box is 11 430 Ω.

The true voltage

$$V = \frac{V_o \times R_x}{R_i} = \frac{6.25 \times 11.430}{10\ 000} = 7.14 \text{ V}$$

Equivalent Resistor Method.   Instead of the variable resistor, a resistor with the same value as the voltmeter at the selected range (R_e) can be used. This method does *not* give a theoretically correct reading, since the formula given is an abbreviation of the correct (and complicated) one. However, the accuracy of the abbreviated formula, if not used to correct loading effects in excess of 10%, is good, generally not more than the instrument's own error.

To use this method:

1. Measure the voltage carefully.
2. Insert R_e in *series* with the meter.
3. Read the voltage, carefully, but remember, that the scale factor is now doubled!
4. Calculate V, the true voltage from the formula

$$V = \frac{V_e^2}{V_o}$$

where V_o is the original voltage reading and V_e is the voltage reading with R_e inserted in the measuring circuit.

Example 3.9

Using the same circuit as the one used in Example 3.11, the voltage measured is $V_o = 6.25$ V. Since R_i is 10 kΩ, a 10-kΩ resistor is inserted in series with the meter (Figure 3.13). The voltage reading (on the doubled scale!) is 6.67 V (V_e), so

$$V = \frac{6.67^2}{6.25} = 7.12 \text{ V}$$

The difference between the two methods,

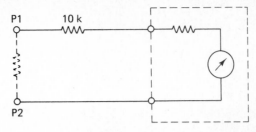

Figure 3.13

$$D = \frac{100 \times 0.02}{7.14} = 0.3\%$$

on the border of the readability of the instrument.

> *Note:* For a rough estimation, the voltage reading with R_e in parallel can be taken and V calculated as V_o^2/V_e. The error of this method is much larger. In the example above it would give a voltage $V = 7.03$ V.

A voltage measurement without any loading effect in any circuit can be made using a *potentiometric instrument*, described in Chapter 7.

3.4.2 Measuring the Voltage of a Source

Unless a source is infinitely strong, *any load will drop the open-circuit voltage*. The voltage drop increases as the load increases (that is, the circuit resistance decreases) and the source strength decreases. The "strength" of the source depends on the size and type of the generator, power supply, output circuitry, and so on.

To be able to *express, measure,* and *calculate* the source strength and its effects, it is expressed in terms of *resistance*, which is the *internal resistance of the source* or the *output impedance of the source*. It can be visualized that the actual source is an infinitely strong source and a resistor (R_s) in series.

> *Note:* The internal resistance of a source is the *virtual* resistance when producing an electromotive force (EMF). It can be measured by its effects only. Do not attempt to hook up an ohmmeter to a source!

R_s, the source output impedance, can be handled and calculated like any other resistance. The two circuits shown in Figure 3.14 are practically identical as long as R_s (the source output impedance) is equal to R_L (the externally added load resistance). To take the loading effect of the source into account for calculations, R_s is assumed to be part of the external circuit resistance.

All the examples discussed in the preceding section are valid for a practical (that is, not infinitely strong) source, provided that R_A, the resistance at the nonmeasured part of the circuit, incorporated R_s, the source resistance.

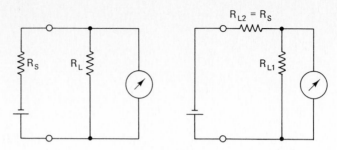

Figure 3.14

Since the effect of source loading is similar in dc and ac circuits and sources, it is called source impedance rather than resistance. Accordingly, the strength of a source is given as the *output impedance*, which is the equivalent internal resistance (for dc) or impedance (for ac) of a source.

Similarly, a circuit to be connected to a source can be expressed by the resistance or impedance of the circuit between the two points of connection to the source. The ratio of this *input impedance* to the output impedance of the source will determine the loading effect, that is, the drop of the supply voltage when the circuit (load) is connected.

The *voltage drop* under load can be calculated in the same way as it was calculated in the previous examples. Since the same current goes through the load and the source (R_s), the equation

$$I = \frac{V_A}{R_A} = \frac{V_B}{R_B}$$

can be used, where A and B can be any selected part (or the total) circuit.

Example 3.10

A car battery has an open-circuit voltage of 12.5 V and an output impedance (R_s) of 0.05 Ω. A 100-W lamp is switched on. How much will that drop the battery voltage?

The current used by the lamp is $A = W/V = 100/12.5 = 8$ A. The resistance of the bulb is $R = V/I = 12.5/8 = 1.56$ Ω. The circuit is shown in Figure 3.15.

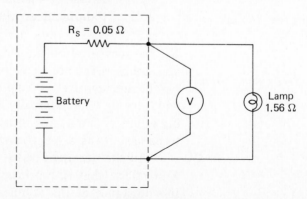

Figure 3.15

The measured voltage is calculated as follows:

$$\frac{12.5\,\text{V}}{(1.56 + 0.05)\,\Omega} = \frac{V_x}{1.56\,\Omega}$$

$$\underset{\substack{\uparrow \\ R\ \&\ V\ \text{of the} \\ \text{full circuit}}}{} \qquad \underset{\substack{\uparrow \\ R\ \&\ V\ \text{drop} \\ \text{across the bulb}}}{}$$

$$V_x = \frac{1.56 \times 12.5}{1.56 + 0.05} = 12.1$$

$$\underset{\substack{\uparrow \\ \text{voltage across the lamp}}}{}$$

There is a 0.4-V voltage drop when the lamp is turned on.

Example 3.11

A temperature-measuring element has an EMF output and an output impedance of 20 000 Ω. The input impedance of the receiving circuit (load) is 1 MΩ. How much error will be introduced by loading the source if the open-circuit EMF is 76.1 mV?

V_x, the voltage "seen" by the receiving circuit, is calculated as follows:

$$\frac{76.1\,\text{mV}}{1.02\,\text{M}\Omega} = \frac{V_x}{1\,\text{M}\Omega}$$

$$V_x = \frac{1 \times 76.1}{1.02} = 74.6\,\text{mV}$$

and the error

$$E = \frac{100\,(76.1 - 74.6)}{76.1} = 2\%$$

This calculation is typical. Any time two circuits are joined when one supplies the other, the loading error can be calculated this way knowing the output and input impedances. The error will be less as the source output impedance *decreases* or the load input impedance *increases*.

3.4.2.1 *Correction of the Loading-Effect Error*

The input impedance of the load can be established by calculation or direct measurement. The output impedance of the source cannot be measured directly, but it can be calculated using the method described in this section by attaching a known resistance to the load and measuring the voltage drop.

The very same thing happens, involuntarily, if the output voltage of a source is measured with an instrument that has less than infinite impedance. That is, the meter itself loads down the source, and the voltage indicated on the voltmeter will *not* be the open-circuit voltage. Figure 3.16 shows the combination of a source with

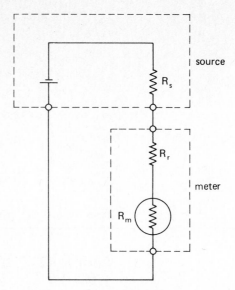

Figure 3.16

R_s output impedance and a voltmeter. *To find the loading error, and to compensate for it, the same methods can be used as were used for the measurement of a voltage across a component.* That is, attach a resistor with a resistance equal to the voltmeter *in parallel* with it and watch the meter needle. If its movement is smaller than the permissible error, accept the reading. If it is larger than that but not more than about 10% of the meter range, compensate for the reading error; if it is even larger, a more sensitive voltmeter must be used.

Compensation for the error is done similarly; that is:

1. With the variable (decade) resistor in *series* with the voltmeter, take a reading with a 0-Ω setting and adjust the resistance setting until the reading of the meter is exactly half of the original indication and calculate V from the formula

$$V = \frac{V_o \times R_x}{R_i}$$

(using the same notation as before), or

2. Attach a resistor (R_e) of equivalent value to the voltmeter in series with it, read the voltage (on the doubled scale!) and calculate the approximate voltage using the equation

$$V = \frac{V_e^2}{V_o}$$

Once the true voltage (or the approximate true voltage) is known, the output impedance of the source can be calculated knowing the true voltage (V), the measured voltage (V_o), and the meter resistance

(R_i), again using the voltage divider formula:

$$(I =) \frac{V_o}{R_i} = \frac{V - V_o}{R_s}$$

(I is the current through the resistances R_1 and R_2) and so

$$R_s = R_i \frac{V - V_o}{V_o}$$

(The measurements obtained with the variable or equivalent resistor are *not* used in this calculation.)

Example 3.12

Measuring a voltage on the range 0 to 5 V with an instrument using a 1-mA f.s.d. movement, the reading is V_o = 3.26 V.

To check the loading effect, an equivalent resistor is put in parallel with R_i, which for the 5-V range is 5/0.001 = 5000 Ω, so R_e will be 5000 Ω. The new reading with R_e in series with the meter, V_e, is 3.44 V. The true voltage

$$V = \frac{V_e^2}{V_o} = \frac{3.44^2}{3.26} = 3.63 \text{ V}$$

The source output impedance

$$R_s = R_i \frac{V - V_o}{V_o} = 5000 \frac{3.63 - 3.26}{3.26} = 567 \ \Omega$$

which is a good approximation.

Example 3.13

Measuring V with the decade box, the resistance on the box to drop the voltage to half (1.63) is 5600 Ω. The true voltage

$$V = \frac{3.26 \times 5600}{5000} = 3.65 \text{ V}$$

and

$$R_s = 5000 \frac{3.65 - 3.26}{3.26} = 600 \ \Omega$$

(To drop I_m to half, that is, cause the reading to halve, the total loop resistance must be doubled; that is, the 5600 Ω on the decade box equals the previous total loop resistance, of which 5000 is R_i, so R_s = 5600 - 5000 = 600 Ω, which can be calculated this way without even knowing V the open-circuit voltage.)

The difference of two similar voltages is used in the calculation for R_s ($V - V_o$). The accuracy of the calculation depends on the accuracy with which these voltages were obtained. An accurate and easily readable instrument is required for this in addition to careful, precise work. Watch that the accuracy with which R_s is given does *not* overstate the accuracy to be expected based on the quality of the input.

Accordingly, in the example calculated by the approximate method, R_s could be given as 570 Ω or, knowing that the calculation for V gives a somewhat *low* figure, as ≈ 600 (the $\approx$ sign indicates that the accuracy is determined by the one significant digit).

Calculation of the true voltage is similar if the output impedance and input impedance are known. In this case no auxiliary measurement with R_e is necessary.

Example 3.14

The output signal of a transmitter has an output impedance of 5.6 kΩ and the input impedance of the receiving instrument is 135 kΩ. What is the true voltage and the error before the compensation if the receiving instrument reads 1.28 V?

Since

$$\frac{1.28 \text{ V}}{135 \text{ k}\Omega} = \frac{x \text{ V}}{140.6 \text{ k}\Omega}$$

$$x = \frac{140.6 \times 1.28}{135} = 1.33 \text{ V}$$

and the error

$$E = \frac{100(1.33 - 1.28)}{1.33} = 4\%$$

Too high. What should be the input impedance of the receiving instrument to keep the error to less than 1%?

We use the same equation, rearranged:

$$1\% = \frac{100(1.33 - x)}{1.33}$$

$$x = 1.33 - \frac{1.33}{100} = 1.32$$

and

$$\frac{1.33 - 1.32}{5.6 \text{ k}\Omega} = \frac{1.32}{x \text{ k}\Omega}$$

$$x = \frac{5.6 \times 1.32}{0.01} = 740 \text{ k}\Omega$$

This proves the accuracy of the calculation, since all equations used are linear, and four times the input impedance causes the error to be one-fourth of the original.

3.5 PROBLEMS   1.   Calculate R_r for the following voltmeters.

	Measured Range	*PMMC Movement*	
(a)	0–10 V	1 mA	200 Ω
(b)	0–150 mV	0.5 mA	300 Ω
(c)	0–2 V	50 μA	250 Ω

2.   Make a sketch and calculate all the resistors (including the adjustable resistor) for the following multirange voltmeters.

	Ranges (V)	*PMMC Movement*	
(a)	0.2		
	1	1.5 mA	100 Ω
	10		
(b)	0.2		
	0.5		
	1	1 mA	150 Ω
	2		
(c)	1		
	5		
	10	0.5 mA	200 Ω
	20		
(d)	0.5		
	2		
	5	2 mA	250 Ω
	20		

3.   Predict the voltage measurement error in the following circuits. The circuit is shown in Figure 3-8, and the voltage is measured across R_B.

	V	R_A (Ω)	R_B (Ω)	I_m	*Meter Range* (V)
(a)	73.2	5000	12 000	0.5 mA	0–100
(b)	0.93	72	150	0.2 mA	0–2
(c)	3.05	25 000	20 000	1 mA	0–5
(d)	12.75	3750	210	50 μA	0–15

4.   Calculate the series resistor R_e and the approximate true voltage if the original voltage and the voltage measured with R_e in series with the meter are.

	V_o	V_e	I_m (mA)	*Meter Range* (V)
(a)	1.945	2.038	1	0–5
(b)	0.566	0.606	0.1	0–1
(c)	40.2	42.1	2	0–50
(d)	16.00	16.76	0.5	0–25

5.   Calculate the error of the voltage reading sensed by the receiving instrument in the following output impedance/input impedance combinations.

	Output Impedance (Ω)	*Input Impedance*	*Voltage Measured* (V)
(a)	75	3000 Ω	2.3
(b)	490	10 200 Ω	0.135
(c)	6500	200 000 Ω	240
(d)	23 000	2.5 MΩ	1.06

6. Calculate the source output impedance if the originally measured
    voltage (V_o) with the meter only and the voltage measured with
    R_e in series with the meter (V_e) is known.

	V_o	V_e	I_m (mA)	Meter Range (V)
(a)	6.22	6.53	10	10
(b)	2.69	2.75	2	5
(c)	54.1	57.2	1	100
(d)	0.97	1.00	0.1	2

4

AC Voltage Measurements

A meter movement using a permanent magnet responds only to dc. (Instruments using an electromagnet in place of the permanent magnet are used for some ac measurements, but they will not be discussed here.)

To enable a PMMC movement to measure ac, the *part of the* current going into the movement must be rectified. The rectification can be *full wave* (Figure 4.1A) or *half wave* (Figure 4.1B). Diode b in Figure 4.1B does not take part in the rectification; it just prevents the reverse current from going through diode a. It allows the meter to present a constant load to the circuit under test, and prevents any reverse current from going through the meter.

The sinusoidal current rectified by a bridge (full-wave) rectifier looks as shown in Figure 4.2. Assuming no phase difference (power factor = 1), both the voltage and current curves will be as shown in Figure 4.2. There are three voltages shown: V_{ave}, V_{rms}, and V.

The deflection of the pointer is proportional to the arithmetic average of the sinusoidal voltage. The ratio of the average to the peak voltage can be calculated, as the area under the sine curve (crosshatched) is made equal to the rectangle crosshatched in Figure 4.2. The horizontal distance is π (180°) and

$$V_{ave} \times \pi = \int_0^\pi \sin V \, d\alpha = [-\cos V]_0^\pi = 1 - (-1) = 2$$

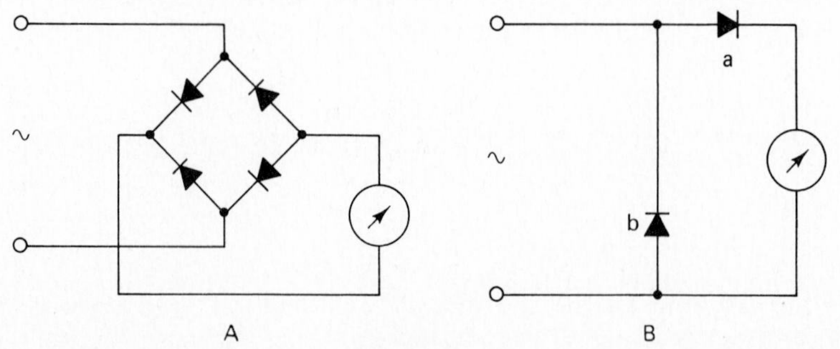

A

B

Figure 4.1

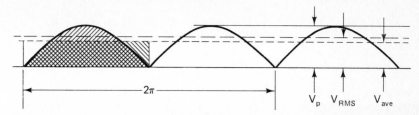

Figure 4.2

and

$$V_{\text{ave}} = \frac{2}{\pi} = 0.637\,(0.63662) \text{ of } V_p$$

Sinusoidal ac is *not* measured by its average value. Ultimately, electricity is expected to perform work, deliver energy. This depends on the W used, and since the power depends linearly on the product of A and V, the *root-mean-square* (rms) value of the V_p is a better approximation of the power-delivering ability of the ac supply. Whether or not this explanation relates to the purpose for which the ac current is used in any given application, *the voltage of any ac is always given as the rms value*, unless specifically stated otherwise. This means the the *voltage scale* of the ac voltage-measuring instrument must be graduated in rms volt units.

According to the definition of rms, the peak-to-peak voltage, $V_{\text{p-p}}$ is $\sqrt{2}$ times V_{rms}. This gives a ratio between V_{ave} and V_{rms}. Since

$$V_{\text{p-p}} = \sqrt{2}\, V_{\text{rms}} = \frac{\pi}{2}\, V_{\text{ave}}$$

$$V_{\text{ave}} = \frac{2\sqrt{2}}{\pi}\, V_{\text{rms}} = 0.9 V_{\text{rms}}\,(0.900316 V_{\text{rms}})$$

$$V_{\text{rms}} = \frac{\pi}{2\sqrt{2}}\, V_{\text{ave}} = 1.11 V_{\text{ave}}\,(1.11072 V_{\text{ave}})$$

using a full-wave rectifier. The numbers in parentheses are for high-accuracy calculations, but for regular measurements the figures 0.9 and 1.11 (which are both accurate to three significant digits!) are used. Since the PMMC movement inevitably shows V_{ave}, and the scale, normally, is in V_{rms}, an allowance must be made for the difference. From the calculations above it can be seen that if the scale of the meter is 1 V_{rms}, only $0.9 V_{\text{ave}}$ has to be supplied to deflect the meter fully. *That is, in practice, the nominal full-scale rms voltage must be multiplied by 0.9 to find the voltage used for R_r calculations.* Most meters have general scales, but the *scale factor* for ac voltages is based on rms voltage, so there is a need to apply the proper multiplying factor, called the form factor ($V_{\text{ave}}/V_{\text{rms}}$).

When using a *single-diode (half-wave) rectifier*, the average voltage will be exactly half that produced by a full-wave rectifier, and the multiplying factor to calculate R_r will be *half as well, 0.45.*

There are a few considerations when using rectified ac. Recti-

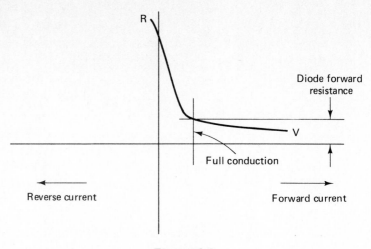

Figure 4.3

fication is done using diodes (old instruments may have copper oxide or selenium rectifiers). These diodes have practically infinite resistance in the reverse direction, but in the forward direction their resistance is *not* negligible. Based on type, size, and voltage rating, their forward resistance can be from several ohms to several hundred ohms. The resistance is not constant; it is *larger* at lower voltages (Figure 4.3). The resistance is caused by the fact that silicon diodes do not conduct below about 0.7 V and germanium diodes below about 0.3 V, depending on the diode and on the temperature.

To get the required accuracy, the diode must be in full conduction; that is, the current through it must be above a minimum value and the effect of the nonconducting voltage minimum taken into account.

There are several ways to achieve this. Using a half-wave rectifier raises the ratio of V_p (and the maximum forward current) to V_{ave} to 3.14 (π), and the initial (high resistance) region of the diode is passed faster because of the higher ac voltage required.

A shunt (resistor in parallel with the meter) increases the current through the diode since it has to provide current for the resistor and the meter, achieving the same result. Figure 4.4 shows the arrangement used for many commercial instruments.

Both methods, in effect, *desensitize* the meter movement. This prevents the meter from being used to measure low ac ranges, increases the loading effect, and decreases the accuracy at higher ranges.

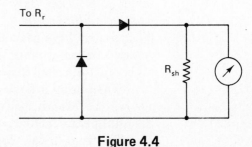

Figure 4.4

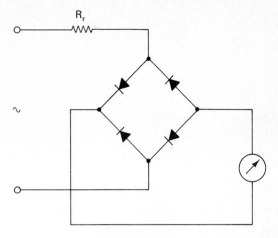

Figure 4.5

Fortunately, modern diodes have better characteristics, and the threshold of full conduction is low enough to assure adequate operation with movements of normal sensitivity (1 mA f.s.d. current).

One way to compensate for the effects of the diode in the circuit is to add the equivalent diode dc resistance (R_d) to the meter resistance when making calculations. The problem is that R_d changes with the voltage applied. In practice, the easiest solution is to calculate the average voltage excluding the nonconducting region (0.7 V for silicon diodes, 0.3 V for germanium diodes) and divide this average voltage by the rms voltage to get the multiplying factor (form factor). Figure 4.5 contains graphs with the multiplying factors for full-wave and half-wave rectifiers to be used for silicon and germanium diodes.

4.2 AC VOLTMETERS Figure 4.6 shows a voltmeter for using a full-wave rectifier.

Example 4.1

Find R_r if the meter used is 1 mA, 200 Ω; the diodes are silicon; and the voltmeter range 0 to 6 V ac. Converting the 6 V (which is rms) into the average for 6 V (from the graph in Fig. 4.5), we obtain

$$V_{ave} = 0.69 \times 6 = 4.14 \text{ V}$$

Thus

$$R = \frac{6 \times 0.69}{0.001} = 4140 \text{ }\Omega$$

$$R_r = R - R_m = 4140 - 200 = 3940 \text{ }\Omega$$

Figure 4.7 shows the arrangement with a half-wave rectifier.

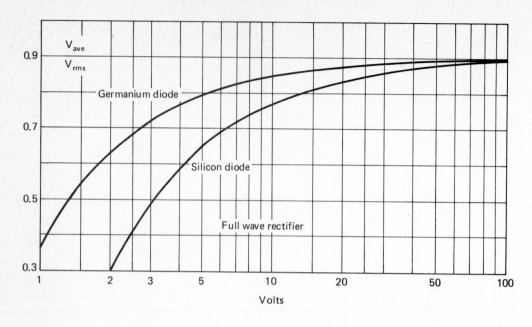

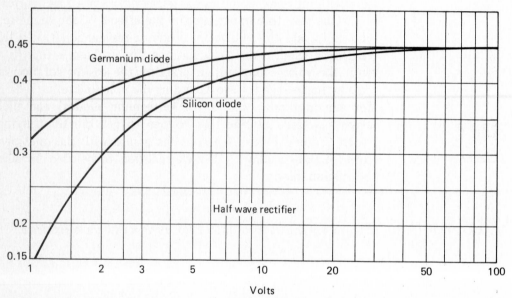

Figure 4.6

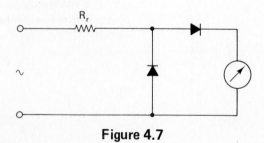

Figure 4.7

Example 4.2

Find R_r if the meter is 0.5 mA, 100 Ω; the diode is germanium; and the range to be measured is 0 to 1 V.

$$V_{\text{ave}} = 0.33 \times 1 = 0.33 \text{ V}$$

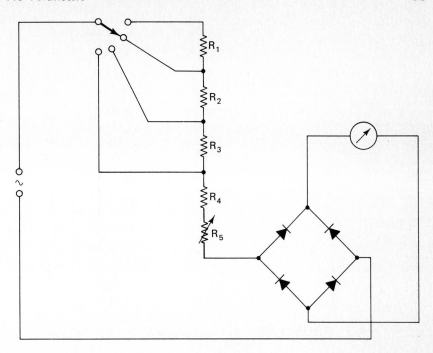

Figure 4.8

Thus

$$R = \frac{0.33}{0.0005} = 660 \ \Omega$$

$$R_r = 660 - 100 = 560 \ \Omega$$

When constructing a multirange ac voltmeter, the same procedure is used. It differs from the dc multirange calculations by finding V_{ave} and including R_d.

Figure 4.8 shows a multirange ac voltmeter with a full-wave rectifier.

Example 4.3

Calculate the resistances if the meter is 1 mA, 150 Ω; the diodes are silicon; and the ranges to be measured are 2, 5, 10, and 20 V.

	Range (V)			
	2	*5*	*10*	*20*
Form factor	0.3	0.645	0.77	0.83
V_{ave} (V)	0.6	3.22	7.7	16.6
R (Ω)	600	3220	7700	16 600
R_r (Ω)	450	3070	7550	16 450
ΔR (Ω)		2620	4480	8900

$$R_4 + R_5 = 450 \ \Omega$$

which can be composed of a 400-Ω resistor and a 100-Ω potentiometer.

$$R_1 = 8900 \ \Omega$$

$$R_2 = 4480 \ \Omega$$

$$R_3 = 2670 \ \Omega$$

$$R_4 = 400 \ \Omega$$

$$R_5 = 100 \ \Omega$$

Warning: When measuring the open-circuit voltage of an ac source, a small load (large R) should be put parallel to the meter because if the meter has any *capacitance*, it may charge the circuit up to V_p. The same applies if a *rectified, but unfiltered* source is measured.

4.3   PROBLEMS   1. Calculate R_r for the following ac voltmeters.

	Range (V)	I_m (mA)	R_m (Ω)	Rectifier
(a)	0–1.5	1	333	Full-wave, germanium
(b)	0–30	2	150	Half-wave, silicon
(c)	0–100	10	100	Full-wave, silicon
(d)	0–500	1	200	Half-wave, silicon

2. Calculate the resistors for the following ac voltmeters. Sketch the circuit.

	Ranges (V)	I_m (mA)	R_m (Ω)	Rectifier
(a)	10, 100	2	200	Full-wave, silicon
(b)	2, 5, 20	1	100	Half-wave, germanium
(c)	2, 5, 10, 50	0.5	150	Full-wave, germanium
(d)	2, 10, 50, 100	1.5	250	Half-wave, silicon

5

Current Measurement

5.1 DC CURRENT MEASUREMENT

The measurement of current gives more problems in practice than the measurement of voltage. One of the reasons is that the current-measuring instrument (ammeter) must be placed *in series* with the circuit. The circuit has to be opened up and new connections made, all of which must pass the current safely.

The output of a circuit or instrument may be a current signal. Current signals, unlike voltage signals, require a definite impedance in the receiving circuitry (which can be a given value or a named high and low limit). This current signal must be measured with the required resistance in the measuring circuit (dummy load). Since ammeters have relatively low resistances, an error may cause blown fuses or burn components. Make sure that all current measurements are well planned, well understood, and that all connections are made with proper care.

5.1.1 Shunt Resistance Calculations

A PMMC movement measures current. Its sensitivity shows the *lowest* current range that the instrument can measure, I_m. Higher current ranges can be measured if a resistor *in parallel* with the meter movement will branch some of the current off. The name of this resistor is *shunt* or *shunt resistor* (R_{sh}). The shunt can be calculated using Kirchhoff's laws. In Figure 5.1 the total current I is branched off into I_1 and I_2 at node point A. Kirchhoff's first law states that the *sum of the branch currents equals the total current*, in this case, $I = I_1 + I_2$.

A and B are common points, and the voltage drop measured between A and B is the voltage drop across both resistors, R_1 and R_2. From Ohm's law, $V = IR$, this can be written to both branches,

$$I_1 R_1 = I_2 R_2$$

because V is common to both, and

$$\frac{I_1}{I_2} = \frac{R_2}{R_1}$$

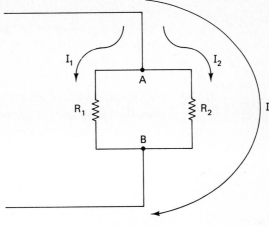

Figure 5.1

which is Kirchhoff's second law: The ratio of the currents is the inverse of the ratio of the resistances. This equation can be written as

$$R_1 = R_2 \frac{I_2}{I_1}$$

It will be used in this form for shunt calculations.

Example 5.1

A 1-mA 100-Ω meter movement is used to measure 5 A (full scale). Calculate R_{sh}. The 5 A will be branched into two parts, I_m and I_{sh} (Figure 5.2). We know that the total current, $I_{sh} + I_m$, is 5 A, and that I_m, the meter current for full deflection, is 1 mA. The current going through the shunt, I_{sh}, can be calculated from Kirchhoff's first law,

$$I_{sh} = I - I_m = 5000 - 1 = 4999 \text{ mA}$$

We have all the data we need to calculate R_{sh} from Kirchhoff's second law:

$$R_{sh} = R_m \ \frac{I_m}{I_{sh}} = 100 \ \frac{1}{4999} = 0.02 \ \Omega$$

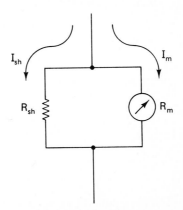

Figure 5.2

The calculator value, $0.020\ 004\ \Omega$, is used if the value of the manufactured R_{sh} must be accurate to at least five significant digits (this surpasses the accuracy and readability of most PMMC instruments). When calculating shunts for regular meter movements, three significant digits are necessary and satisfactory. For high-accuracy meter movements four significant digits are correct, but only if R_{sh} can be manufactured and/or *measured* to that accuracy.

Note also that the accuracy of the calculation depends on the *exact* value of R_m. This must be measured accurately rather than taken over from the specifications.

Example 5.2

Calculate R_{sh} if the movement is 0.1 mA and 196 Ω. The current range to be measured is 0 to 800 mA.

$$I_{sh} = 800 - 0.1 = 799.9\ \text{mA}$$

$$R_{sh} = 196\ \frac{0.1}{799.9} = 0.0245\ \Omega$$

Example 5.3

Calculate R_{sh} if a current about 1.2 A is measured with a PMMC movement of 1.5 mA and 103.5 Ω.

The chosen full-scale current range will be 2 A.

$$I_{sh} = 2000 - 1.5 = 1998.5\ \text{mA}$$

$$R_{sh} = 103.5\ \frac{1.5}{1998.5} = 0.0777\ \Omega$$

(The answer $0.077\ 68\ \Omega$ is acceptable if conditions warrant it.)

5.1.2 Multirange Ammeters

A multirange ammeter can be constructed using a switch and the required number of shunt resistors, as Figure 5.3 shows. This arrange-

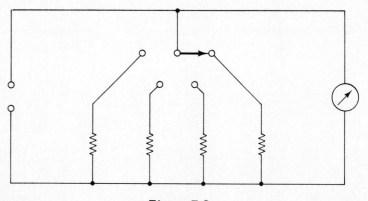

Figure 5.3

ment has one fatal (to the movement) drawback: As the range is switched, the contact is broken momentarily with every shunt, and the movement gets the full amperage. One way to avoid this is to use a make-before-brake type of switch; another way is to use an Arytron shunt.

The make-before-break switch still has the possibility of malfunction, dirt, and broken leads, which will ruin the movement. Multirange ammeters are therefore, of the Arytron type. Figure 5.4 shows the construction of a four-range Arytron shunt. All resistors are part of the circuit at all ranges. When the switch is in the "1" position, $R_1 + R_2 + R_3 + R_4$ constitute the shunt. This is the highest resistance, thus the lowest current range. When the switch is in the "2" position, $R_2 + R_3 + R_4$ make the shunt, while R_1 acts as a ranging resistor in series with the meter. In the "3" position, $R_3 + R_4$ is the shunt, and $R_1 + R_2$ serve as ranging resistors. Finally, in the "4" position, R_4 is the only shunt resistor; the rest are in series with the meter (the range for the highest current). The calculation of the resistors is based on the basic equation written for each switch position, and the solution of these simultaneous equations.

If we designate the current that goes through the meter and the resistors in series as I_m and the current going through the shunt resistor(s) as I_{sh}, the equations for the circuit shown in Figure 5.4 will be as follows:

1. $I_{sh}(R_1 + R_2 + R_3 + R_4) = I_m R_m$ since R_{sh} is $R_1 + R_2 + R_3 + R_4$. This equation is not solved for R_{sh} this time.
2. $I_{sh}(R_2 + R_3 + R_4) = I_m(R_m + R_1)$
3. $I_{sh}(R_3 + R_4) - I_m(R_m + R_1 + R_2)$
4. $I_{sh}R_4 = I_m(R_m + R_1 + R_2 + R_3)$

In all these equations $I_{sh} = I - I_m$, where I is the total current being measured.

There are several ways to solve the simultaneous equations. For programmable calculators, the use of matrices can be recommended.

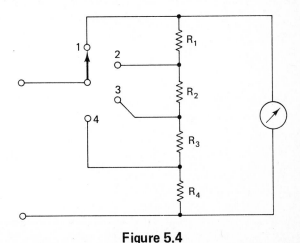

Figure 5.4

Only one way will be shown here. No matter which method used, the only possible problem is loss of accuracy. Since at one point or another, the numbers are added or subtracted, enough resolution must be carried to provide the required accuracy for each calculated value. (Carrying a high number of significant digits at an intermediate calculation does not violate the accuracy laws as long as the answer is correctly given.)

The solution of the equations is shown by several examples.

Example 5.4

Construct a multirange ammeter with ranges of 0 to 0.1, 1, 5, and 10 A using a 1-mA 100-Ω meter movement. The basic equations are as follows:

1. $99(R_1 + R_2 + R_3 + R_4) = 1 \times 100$ since $I_{sh} = 100 - 1 = 99$. All currents are given in milliamperes.
2. $999(R_2 + R_3 + R_4) = 1(100 + R_1)$
3. $4999(R_3 + R_4) = 1(100 + R_1 + R_2)$
4. $9999R_4 = 1(100 + R_1 + R_2 + R_3)$

After the multiplication the equations are rearranged to bring all terms containing unknown R values to the left side. The equations will become:

1. $R_1 + R_2 + R_3 + R_4 = \dfrac{1 \times 100}{99} = 1.0101$
2. $-R_1 + 999R_2 + 999R_3 + 999R_4 = 100$
3. $-R_1 - R_2 + 4999R_3 + 4999R_4 = 100$
4. $-R_1 - R_2 - R_3 + 9999R_4 = 100$

(a) Add equations 1 and 4.

$$\begin{array}{r} +R_1 + R_2 + R_3 + R_4 = 1.0101 \\ + \quad -R_1 - R_2 - R_3 + 9999R_4 = 100 \\ \hline 10\,000R_4 = 101.01 \end{array}$$

$$R_4 = \frac{101.01}{10\,000} = 0.0101 \ \Omega$$

(b) Subtract equation 4 from equation 3.

$$\begin{array}{r} -R_1 - R_2 + 4999R_3 + 4999R_4 = 100 \\ - \quad +R_1 + R_2 + R_3 - 9999R_4 = -100 \\ \hline +5000R_3 - 5000R_4 = 0 \end{array}$$

$$R_3 = R_4 = 0.0101 \ \Omega$$

(c)   Subtract equation 3 from equation 2.

$$-R_1 + 999R_2 + 999R_3 + 999R_4 = 100$$
$$- \quad \frac{+R_1 + R_2 - 4999R_3 - 4999R_4 = -100}{1000R_2 - 4000R_3 - 4000R_4 = 0}$$

Substitute the known values for R_3 and R_4.

$$1000R_2 = 4000 \times 0.0101 + 4000 \times 0.0101$$

$$R_2 = \frac{2 \times 40.4}{1000} = 0.0808$$

(If there were more ranges, the solution would continue by subtracting the equation last handled from the preceding one, and so on).

(d)   Calculate R_1 from equation 1.

$$R_1 = 1.0101 - R_2 - R_3 - R_4$$

$$= 1.0101 - 0.0808 - 0.0101 - 0.0101$$

$$= 0.909 \; \Omega$$

Notice that to get all final answers to three-digit accuracy, the right side of equation 1 had to be given to five significant digits.

Example 5.5

Construct a multirange ammeter for the ranges 0 to 0.05 and 0.5 and 5 A using a 2.5-mA 150-Ω meter movement. The equations are:

1. $47.5(R_1 + R_2 + R_3) = 2.5 \times 150$
2. $497.5(R_2 + R_3) = 2.5(150 + R_1)$
3. $4997.5R_3 = 2.5(150 + R_1 + R_2)$

Rearranged, they are as follows:

1. $R_1 + R_2 + R_3 = \dfrac{2.5 \times 150}{47.5} = 7.8947$

Since in the following equations the resistances on the right side will have a multiplier of 2.5, to facilitate further calculations, equation 1 will be multiplied by this factor:

1a. $2.5R_1 + 2.5R_2 + 2.5R_3 = 19.73$
2. $-2.5R_1 + 497.5R_2 + 497.5R_3 = 375$
3. $-2.5R_1 - 2.5R_2 + 4997.5R_3 = 375$

Adding equations 1a and 3, we have

$$5000R_3 = 394.737$$

$$R_3 = 0.0789 \ \Omega$$

Subtracting equation 3 from equation 2, we obtain

$$500R_2 = 4500R_3$$

$$R_2 = 9R_3 = 9 \times 0.0789 = 0.710 \ \Omega$$

and

$$R_1 = 7.8947 - 0.710 - 0.0789 = 7.106 \ \Omega \quad \text{(or 7.11 } \Omega\text{)}$$

5.2 AMMETER LOADING EFFECT

An ideal ammeter has zero resistance (just as an ideal voltmeter has infinite resistance). Since this is impossible, the resistance introduced into the circuit with the meter will increase the total circuit resistance, lowering the current that was to be measured. The loading effect will be shown in the following example.

Example 5.6

Measure the current in the loop shown in Figure 5.5 (R_A is the *combined* resistance of all loop components, *including* the source output impedance). V is 12 V, $R_A = 76.0 \ \Omega$, the loop current $I_A = 12/76 = 157.9$ mA, the ammeter range will be chosen as 250 mA, and the available PMMC movement is 1 mA, 100 Ω. With this movement $I_{sh} = 249$ and $R_{sh} = 100(1/249) = 0.402 \ \Omega$.

The total resistance of the ammeter will be

$$R = \cfrac{1}{\cfrac{1}{R_m} + \cfrac{1}{R_{sh}}} = \cfrac{1}{\cfrac{1}{100} + \cfrac{1}{0.402}} = 0.4 \ \Omega$$

If this resistance is added to R_A, the loop current will be

$$I_A = \frac{12}{76.4} = 157.1 \ \text{mA}$$

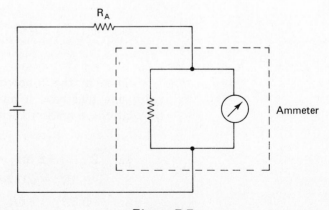

Figure 5.5

and the error

$$E = \frac{100(157.9 - 157.1)}{157.9} = 0.5\%$$

This is not a significant error, but it is not negligible for high-accuracy measurements.

If a more sensitive, 0.1-mA 100-Ω movement were used, we would have

$$R_{sh} = 100 \frac{0.1}{249.9} = 0.04 \ \Omega \qquad \text{and} \qquad R = 0.04 \ \Omega$$

This cuts the current change and the error 10-fold.

It can be concluded that *the higher the meter sensitivity, the lower will be the loading error.* This is the same for current as it was for voltage.

If we use a 1-mA 200-Ω movement in the same application, I_{sh} will be the same, but

$$R_{sh} = 200 \frac{1}{249} = 0.804 \ \Omega$$

So $R = 0.8 \ \Omega$, and the loading and the error will be double. *A lower meter resistance will reduce the loading error* (in contrast to voltage measurements, where the meter resistance did not influence the loading error).

If the original 1-mA 100-Ω movement on the 250-mA range is used to measure the current in a circuit where $V = 1.2$ V and $R_A = 7.6 \ \Omega$, the loading error will be (since the meter resistance is the same)

$$I_A \ \text{original} = \frac{1.2}{7.6} = 157.1 \ \text{mA}$$

$$I_A \ \text{with meter} = \frac{1.2}{8.0} = 150 \ \text{mA}$$

$$E = \frac{100(157.1 - 150)}{157.1} = 4.5\%$$

The figures above indicate that the ratio of the original loop resistance, R_A, to the resistance of the ammeter, R_M (which is the combined parallel resistance of R_m and R_{sh}), determines the size of the loading error).

The loading error can be calculated using the formula

$$E \ (\%) = 100 \left(1 - \frac{R_A}{R_A + R_M}\right)$$

where R_A and R_M are as explained above.

Note: The derivation of this formula is simple, since the original current is $I_A = V/R_A$, and the new current is $I_N = V/(R_A + R_M)$; then the error

$$E = \frac{V/R_A - V/(R_A + R_M)}{V/R_A}$$

This reduced to the formula given above. R_A must include the source output impedance.

For higher currents (10 A and over, depending on the circumstances) specially constructed ammeters are used. The procurement or production of very low resistance shunts is difficult. [For resistances in the range 0.01 to 1 Ω, a carefully trimmed length of resistance or thermocouple wire can be used, based on the known resistance/length value of the wire. For higher currents, though, the heat dissipation of the resistor may not be enough, causing overheating and consequent resistance change. For lower values of resistance the resistance of the leads, binding post, and joints should be known. The *measurement* of the shunt assembly requires special equipment, (a Kelvin bridge, for example) and very careful work.]

5.3 AC CURRENT MEASUREMENT

5.3.1 Single-Range AC Ammeters

The generally low resistance of an ammeter (movement–shunt combination) makes the voltage drop across it too low to put it into the conductive range of a diode. The usual arrangement is shown in Figure 5.6A. This puts the diodes into the circuit of the full current, where the added resistance may be objected. The arrangement in Figure 5.6B corrects that if the voltage drop is sufficient for conduction.

The difference between the average and rms currents (the form factor, which is in the same relation as the average and rms currents) requires consideration in calculating the value of the shunt.

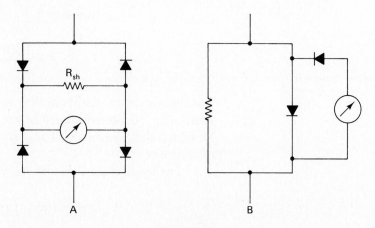

Figure 5.6

Since the average value is less than the rms, a shunt calculated to give full deflection for the meter using the average current will *not* deflect the meter fully. To direct more current to the meter, the value of the shunt must be *increased* by the rms/average ratio.

In practice, the rms current going through the meter–rectifier is *multiplied by 1.11 for full-wave rectifiers and by 2.22 for half-wave rectifiers*. The rest of the calculation, that is, the calculation to find the current going through the shunt and the calculation of the shunt resistance is the same as for the dc ammeters. The resistance of the diode (Fig. 5.6B) adds to the resistance of the instrument when making calculations.

Example 5.7

Use a 0.5-mA 200-Ω movement to make an ammeter to measure 2 A ac with a full-wave rectifier.

$$I_m = 0.5 \text{ mA}$$

Corrected for the average, $1.11 \times 0.5 = 0.555$ mA.

$$I_{sh} = 2000 - 0.555 = 1999.455 \text{ mA}$$

$$R_{sh} = 200 \, \frac{0.555}{1999.455} = 0.0555 \, \Omega$$

The diodes in this circuit do not affect the calculation of R_{sh}, but their influence on the circuit (loading) must be kept in mind.

Example 5.8

A 1-mA 150-Ω movement with a 40-Ω diode in a half-wave rectifier is to be used to measure 0.8 A full scale. Calculate R_{sh}.

$$I_{sh} = 800 - 2.22 \times 1 = 797.79 \text{ mA}$$

$$R_{sh} = (150 + 40) \, \frac{2.22}{797.79} = 0.529 \, \Omega$$

The ac ammeter calculations in this section are limited to *sinusoidal low-frequency* ac, which is in the 60-Hz household current.

If the waveform is not sinusoidal, the value of the average is different. With other than sinusoidal waveforms, the current may be given as average or peak. Conversions must be made based on the geometrical shape of the wave.

At low frequencies the capacitive and the inductive impedance of the components can be neglected and only the ohmic resistances are considered and calculated. With increasing frequency this consideration is less valid. The inductance and capacitance of an ammeter depends largely on the type of components used and on the arrange-

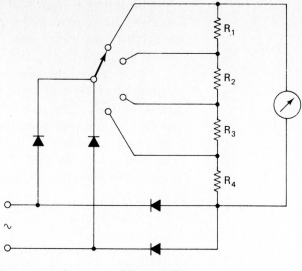

Figure 5.7

ment, interconnection, and housing of the components; therefore, no rules or calculations can be given for frequency limitations, but problems are seldom serious below 1 kHz and never negligible over 100 kHz.

5.3.2 Multirange AC Ammeters

Multirange ac ammeters are constructed in the same way as dc ammeters are constructed. If full rectification is used, the diodes must be properly sized and their effect on the loop taken in account. Figure 5.7 shows such an arrangement.

5.4 PROBLEMS

1. Select the meter range and calculate R_{sh} for the selected meter range for the following current measurements. (R_s is the source output impedance and R_A is the resistance of the circuit.)

	Supply Voltage (dc)	R_s (Ω)	R_A (Ω)	I_m	R_m (Ω)
(a)	0.45	300	1750	50 μA	250
(b)	6	4000	12 000	0.5 mA	150
(c)	24	75	315	1 mA	100
(d)	135	—	5400	1.5 mA	200
(e)	12	750	—	1 mA	300

2. Draw the circuit and calculate the resistors for the following multirange ammeters.

	Range (A)	I_m (mA)	R_m (Ω)
(a)	0-1		
	0-5	1	200
	0-10		

$$(b) \begin{cases} 0\text{-}0.25 \\ 0\text{-}1.0 \\ 0\text{-}5 \\ 0\text{-}10 \end{cases} \qquad 0.1 \qquad 250$$

$$(c) \begin{cases} 0\text{-}1 \\ 0\text{-}5 \\ 0\text{-}10 \\ 0\text{-}20 \end{cases} \qquad 0.5 \qquad 100$$

3. Calculate the loading error for the ammeters used in Problem 1.

4. Select a PMMC meter to measure the current in a loop where $V = 12$ and the sum of the resistances 11 700 Ω to keep the error below 1%.

5. Calculate R_{sh} for the following current measuring instruments.

	Range Amperage (ac)	I_m	R_m (Ω)	Rectifier	R_d (each) (Ω)
(a)	0–015	1 mA	150	Half-wave	15
(b)	0–2	1 μA	350	Full-wave	25
(c)	0–10	0.5 mA	200	Half-wave	40
(d)	0–16	2 mA	100	Full-wave	20

6. Draw the circuit of a multirange ammeter (four ranges) using (a) full-wave and (b) half-wave rectifiers.

6

Resistance Measurement

Voltage and current measurements are made *in an active circuit*. Resistance measurements are made on separate components. For most resistance-measuring methods, the component must not be a part of a circuit that carries current.

The reason for this is that the components (or connected components) to be measured are made part of a measuring circuit, where the resistance of that component determines the voltage or current, which is measured by known methods. If a resistance is unknown, or resistances of different values are measured, a *series or parallel ohmmeter is used*, for high-accuracy measurements, but for limited ranges, *bridges are used*. Series and parallel ohmmeters measure ohmic resistance only. Bridges can measure capacitive and inductive impedance as well.

6.1 SERIES AND PARALLEL OHMMETERS

In a simple current-measuring circuit, such as in Figure 6.1, the current is determined by V and R (Ohm's law), which are not known at the time of the measurement. If one of these variables (V or R) is known, with the measured current the other variable can be calculated (Ohm's law again).

For the calculation of the unknown variable it makes no difference whether the meter is set up as a "voltmeter" (Figure 6.2A) or an "ammeter" (Figure 6.2B).

6.1.1 Series Ohmmeters

If the value of the resistor(s) are known, the voltage can be calculated; that is exactly how R_r is calculated for voltage measurements.

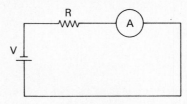

Figure 6.1

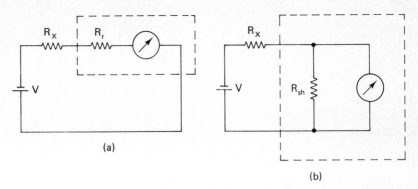

Figure 6.2

If V is known, R can be calculated. More precisely, if V and R_r or V and R_{sh} are known, R_x can be calculated.

First, the bad news: V must be *accurately* known; if its value is changed or not known, the resistance measurement will be changed the same ratio. The good news is that the PMMC movement can be supplied with a *special scale* which shows the resistance of R_x *without* any calculation.

The voltage supply can be adjusted with a variable resistor in series or as part of R_r in the circuit. Including a known value for R_x, the variable resistor can be adjusted to show the correct value on the scale. In practice, $R_x = 0$ is chosen (that is, the meter is shorted), and the variable resistor is adjusted for the movement to show zero on the scale (not necessarily 0 mA!).

The circuit in Figure 6.3 is arranged similar to that of a voltmeter (Figure 6.2A), but arranged to measure resistance. To make discussion of the scale simple, assume that V is known and that there is no loading effect. Let $V = 1$ V and the movement be 1 mA, 200 Ω. The meter is to read the resistance from one extreme to the other, that is, from $R_x = 0$ to $R_x = \infty$.

If $R_x = 0$ (the test points are shorted), *maximum* current goes through the meter. It becomes identical to an ordinary voltmeter and R_r can be calculated as $R_r = V/I_m - R_m = 1/0.001 - 200 = 800$. That is, if $R_r = 800$ Ω, zero ohms measured gives a full milliampere reading on the instrument. At the other extreme, if $R_x = \infty$ (that is, the circuit is open), the meter indication will be 0 mA.

It is easy to see that V and the PMMC movement can be selected to be of any value, and as long as R_r is calculated correctly, at

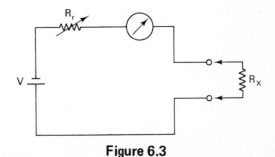

Figure 6.3

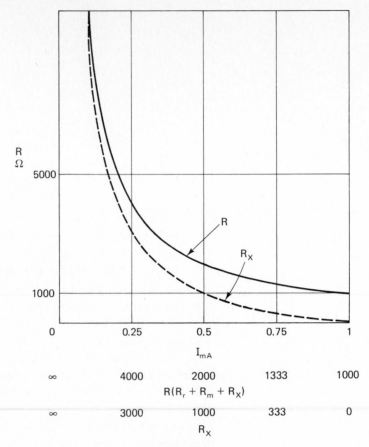

Figure 6.4

$R_x = 0$, the indication will be full meter deflection, and at $R_x = \infty$, zero meter deflection. That is, all series-type voltmeters, regardless of the values of the components used, *measure resistance from zero to infinity, and the $R_x = 0$ reading is at the top of the scale.* The values of R_x between zero and infinity cause different indications, though, depending on the movement and voltage supply used.

Referring to Figure 6.3, the reading of the meter will be midpoint, when the current in the loop is exactly half (from Ohm's law again). That happens when the resistance in the loop is exactly double that of the previous value (provided, again, that the reduced load will not let V rise). This happens when R_x is equal to the rest of the circuit resistance, which is the meter resistance, $R_r + R_m$.

At any other point of the scale, the meter indication, which is the *current* going through the meter, will be $I = V/R$, and as long as V is constant, C, $I = C/R$, where R is the *total* resistance in the circuit, $R_m + R_r + R_x$. It shows that the meter reading (I) varies *inversely* with the resistance in the circuit. If put on a graph, the corresponding I and R values will be on a *hyperbola*; therefore, when the resistance values are put on the movement indicating scale, it will be a hyperbolic scale. Figure 6.4 shows on a graph the current-resistance points calculated by the $I = V/R$ equation using the example values given: that is, $V = 1$ V, $R_r + R_m = 1000$ Ω, and

$R = R_x + R_r + R_m = R_x + 1000$, and the indication on the 0–1 mA scale is

$$I = \frac{1}{1000 + R_x} \qquad \text{or} \qquad I\,(\text{mA}) = \frac{1000}{1000 + R_x}$$

The solid line on the graph shows the correlation between R and I. The dashed line, below it, shows the correlation between I and R_x. This was made by subtracting the constant value of $R_r + R_m$, 1000 Ω in this case, from each point of the measurement. The two scales on the bottom show the values of R and R_x, respectively, for each measured current value.

The current, that is, the meter indication for any point, can be calculated by rearranging the equation given above:

$$R_x = \frac{1}{I} - 1000$$

(I in the equation is in amperes.)

If a scale is to be constructed showing the resistance at every 10% reading (0, 0.1, 0.2 mA, etc.), all values can be calculated from this equation. The scale for our example will be:

Reading (mA)	Resistance (Ω)
0	∞
0.1	9000
0.2	4000
0.3	2333
0.4	1500
0.5	1000
0.6	667
0.7	429
0.8	250
0.9	111
1	0

A scale in round resistance units can be read more easily, but for that the hyperbolic scale must be constructed with nonlinear graduation. Selected points can be calculated, but care must be taken to choose R_x values in a progressively increasing way.

Example 6.1

For the sample values given above, construct a scale by fixing the position on the scale for the following resistance values: 100, 200, 300, 500, 1000, 2000, 3000, 5000, 7500, 10 000, and 50 000 Ω.

The equation is

$$I = \frac{1}{1000 + R_x}$$

which can be converted to scale percent by multiplying it by 100:

$$\text{scale \%} = \frac{100 \times 1000}{1000 + R_x}$$

R_x	Scale %
100	90.9
200	83.3
300	76.9
500	66.7
1000	50.0
2000	33.3
3000	25.0
5000	16.7
7500	11.8
10 000	9.1
50 000	2.0

The two equations can be generalized for any given value of V, R_r, and R_m, and

$$\text{current (mA)} = \frac{V \times 1000}{R_x + R_r + R_m}$$

The scale percent is calculated knowing that

$$\frac{I\,(\text{indicated})}{I_m} = \frac{x\%}{100\%}$$

$$x\% = \frac{100 \times I\,(\text{indicated})}{I_m}$$

and

$$\text{scale \%} = \frac{V \times 100 \times 1000}{I_m\,(R_x + R_r + R_m)} \qquad (I_m \text{ is in milliamperes.})$$

From this equation R_x can be expressed as

$$R_x = \frac{V \times 10^5}{S\% \times I_m} - (R_r + R_m)$$

Since all series ohmmeters measure R_x from zero to infinity, the only measure of the sensitivity is the midscale resistance, which is equal to the resistance of the meter ($R_r + R_m$). The resistance readings at the lower one-third of the scale are too crowded for accurate reading.

If low R_x values are expected, the midscale resistance should be low, to permit good readability on an expanded low end; and for high R_x values the midscale resistance should be high to bring the reading into the midregion of the scale.

The meter resistance is determined by V and I_m. Since V is taken from a dry cell in the meter, for the sake of convenience and

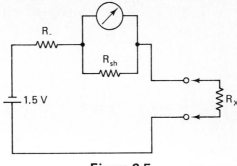

Figure 6.5

safety its value is usually the nominal value of one cell (1.5 V), sometimes of two or four cells (3 or 6 V).

Meter sensitivity for regular (that is, not too expensive) commercial movements is from 50 μA to around 1 mA, which in the combination of the voltages mentioned give a range of midpoint resistance values of roughly 1000 to 100 000 Ω.

The measurement of higher resistances require more sensitive meters; that is, the simple PMMC movement must be substituted by an electronic voltmeter. To measure smaller resistances, the midpoint reading can be lowered if the meter sensitivity is reduced by splitting the current between the movement and a shunt resistor (Figure 6.5). In this case the midpoint resistance is arbitrarily determined.

This resistance is equal to the resistance of the measuring circuit if $R_x = 0$. The current through the circuit is calculated using this resistance and the supply voltage.

If this current is higher than I_m, a shunt is put in parallel with the movement and its value calculated to branch off the excess current.

The combined resistance of the meter and the shunt is calculated and this value is subtracted from the loop resistance given above (= midscale resistance) to give R_r.

The shunt resistance can be an adjustable resistance to adjust exactly the midscale resistance. That is, R_r can be made adjustable to adjust the *zero* and R_{sh} can be made adjustable to adjust the *sensitivity*.

Example 6.2

A 1.5-V source and a 1-mA 100-Ω meter is used to make an ohmmeter to have a 200-Ω midscale reading. Calculate the resistors. The loop will look as shown in Figure 6.5.

With $R_x = 0$, the current $I = V/R = 1.5/200 = 7.5$ mA. This should deflect the meter fully so that 1.0 mA goes through the movement and (according to Kirchhoff's first law) 6.5 mA goes through R_{sh}. From Kirchhoff's second law,

$$R_{\text{sh}} = R_m \frac{I_m}{I_{\text{sh}}} = 100 \frac{1.0}{6.5} = 15.4 \ \Omega$$

The combined resistance of the meter and the shunt is

$$R = \frac{1}{\dfrac{1}{100} + \dfrac{1}{15.4}} = 13.3 \ \Omega$$

and $R_r = 200 - 13.3 = 186.7 \ \Omega$.

Example 6.3

The voltage is 3 V; the instrument is 0.1 mA, 250 Ω; and the intended midscale reading 500 Ω.

The current with $R_x = 0$ is

$$I = \frac{3}{500} = 6 \ \text{mA}$$

The shunt current

$$I_{\text{sh}} = 6 - 0.1 = 5.9 \ \text{mA}$$

The shunt resistance

$$R_{\text{sh}} = 250 \ \frac{0.1}{5.9} = 4.24 \ \Omega$$

The combined resistance

$$R = \frac{1}{\dfrac{1}{250} + \dfrac{1}{4.24}} = 4.17 \ \Omega$$

The ranging resistor

$$R_r = 500 - 4.17 = 495.83 \ \Omega$$

The problem with the method of applying a shunt is the increasing current draw. The lower the midscale resistance, the higher the current draw becomes. So far we assumed the voltage source to be strong enough to prevent any loading effect. This is a fair assumption as long as the current is low. In Examples 6.2 and 6.3, at $R_x = 0$ the current was 7.5 and 6 mA, respectively. This is a permissible load on a fresh, good-quality battery. But reducing the midscale resistance lower—for example, 20 Ω for a 1.5-V source, will give a 75-mA load for $R_x = 0$ and 37.5 mA for a half-scale 20-Ω reading. This will cause a loading-effect error (an error that increases as lower resistances are measured) and a premature demise of the battery.

As the battery gets old, its output impedance, R_s, is increasing. Making R_r and R_{sh} adjustable solves the problem only partially, since the loading error changes with the measured values.

Multirange ohmmeters can be constructed by switching different values of R_r and R_{sh} into the circuit.

Example 6.4

Make a multirange ohmmeter with center-scale resistances of 500, 1000, 10 000, and 60 000 Ω, using a 50-μA 200-Ω movement and a 3-V supply.

	Range (Ω)			
	500	*1000*	*10 000*	*50 000*
Current (mA)	6	3.0	0.3	50
Current through R_{sh} (mA)	5.95	2.95	0.25	—
R_{sh} (Ω)	1.68	3.39	20	—
Resistance of R_{sh} and R_m (Ω)	1.67	3.33	18.2	—
R_r (Ω)	498.3	996.7	9981.8	59 800

The circuit design is shown in Figure 6.6. If a shunt is necessary for some or all of the ranges, a two-gang switch is necessary.

6.1.2 Parallel Ohmmeters

The limitation of the series ohmmeter is its decreasing resistance when measuring low-midpoint-resistance ranges. For these applications the *parallel ohmmeter* gives less loading, therefore better reliability.

Figure 6.7 shows the basic arrangement of a parallel ohmmeter. R_r is calculated the same way as for the series ohmmeter, that is, to give full deflection to the PMMC movement when it receives maximum current. In this arrangement the current will be *maximum*

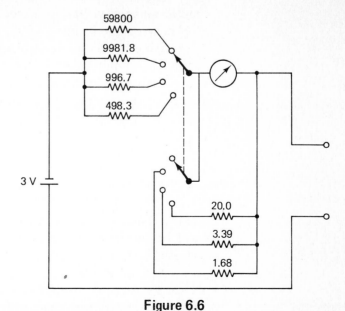

Figure 6.6

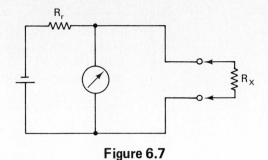

Figure 6.7

when $R_x = \infty$; that is,

$$R_r = \frac{V}{I_m} - R_m$$

When R_x is attached, part of the current goes through R_x and part of it through the movement, reducing the current through the meter and the meter indication. Finally, when $R_x = 0$, all the current bypasses the movement and the meter deflection will be zero.

The parallel ohmmeter, like the series ohmmeter, measures the resistance from zero to infinity regardless of the values of the components used. The major difference is that the *movement indicates zero when $R_x = 0$ and full scale when $R_x = \infty$*. The sensitivity of the parallel ohmmeter is expressed by the resistance measured when the pointer is midscale (same as the series ohmmeter).

It is easy to deduct from Kirchhoff's laws that if $R_x = \infty$ causes full-meter deflection, then $R_x = R_m$ will half the current into both branches and cause exactly a half-scale deflection. That is, the sensitivity of the parallel ohmmeter is determined by the movement resistance, R_m. If, in the circuit shown in Figure 6.7, the meter resistance is 100 Ω, the ohmmeter reading is midscale when $R_x = 100 \ \Omega$.

The meter sensitivity I_m does not play a direct role in the ohmmeter sensitivity, but its effect is not negligible. Lower I_m requires higher circuit resistance and R_r, which will lower the source loading. For saving the battery and minimizing loading error, the movement sensitivity should be as high as possible.

For example, using a 1.5-V battery and a 1-mA 100-Ω movement, the resistance to give 1 mA current with $R_x = \infty$ is

$$R = \frac{V}{I} = \frac{1.5}{0.001} = 1500 \ \Omega$$

and

$$R_r = R - R_m = 1500 - 100 = 1400 \ \Omega$$

This 1400-Ω resistor in series limits the loading of the source, and it will be the loop resistance when $R_x = 0$ (that is, the movement is shorted out).

For comparison: Constructing a series ohmmeter for 100-Ω midscale reading, using the same components, the ohmmeter loop resistance would vary from 100 Ω to infinity as R_x changes from 0 to infinity, and it will be 200 Ω at midscale, thus loading the battery

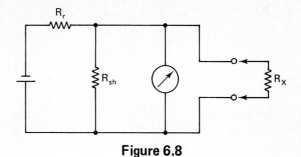

Figure 6.8

about roughly 10 times more at the most used part of the meter. Since the total resistance of the parallel ohmmeter changes from 1400 to 1500 Ω under the same conditions, it will have a 1450-Ω resistance at midscale reading.

There are PMMC movements with lower resistance than 100 Ω (taut band suspension and low-sensitivity meter movements), but the usual way of changing the midscale sensitivity of the ohmmeter is to put a shunt resistor in parallel with the movement (Figure 6.8).

Example 6.5

Construct an ohmmeter using a 1.5-V source and a 0.1-mA 200-Ω movement to have a 50-Ω midscale reading.

Since the combined resistance, R_c, of the movement and R_{sh} is 50 Ω (the midscale reading) and R_m = 200 Ω, R_{sh} can be calculated by rearranging the parallel resistor formula:

$$R_{sh} = \frac{1}{\dfrac{1}{R_c} - \dfrac{1}{R_m}}$$

In this example

$$R_{sh} = \frac{1}{\dfrac{1}{50} - \dfrac{1}{200}} = 66.7 \ \Omega$$

At full-scale deflection, when 0.1 mA current goes through the movement, the current through the shunt will be (Kirchhoff's law)

$$I_{sh} = I_m \frac{R_m}{R_{sh}} = 0.1 \frac{200}{66.7} = 0.3 \ \text{mA}$$

and the total loop current

$$I = I_{sh} + I_m = 0.1 + 0.3 = 0.4 \ \text{mA}$$

The loop resistance can now be calculated to give 0.4 mA:

$$R_e = \frac{V}{I} = \frac{1.5}{0.0004} = 3750 \ \Omega$$

and since 50 Ω is the resistance of the meter–R_{sh} combination,

$$R_r = 3750 - 50 = 3700 \ \Omega$$

Example 6.6

Calculate R_r and R_{sh} for an ohmmeter using a 3-V supply and a 1-mA 100-Ω movement to have 20-Ω midscale reading.

Using the same calculations, we have

$$R_{sh} = \frac{1}{\dfrac{1}{20} - \dfrac{1}{100}} = 25\ \Omega$$

$$I_{sh} = 1\ \frac{100}{25} = 4\ \text{mA}$$

$$R_e = \frac{3}{0.005} = 600\ \Omega$$

$$R_r = 600 - 20 = 580\ \Omega$$

The less sensitive movement and the shunt results in a higher current. The 5-mA load on a good battery is still safe, but it gets worse as the midscale reading is lowered. If, for example, the battery and movement in Example 6.6 were to be used to measure 2 Ω midscale, we would have

$$R_{sh} = \frac{1}{\dfrac{1}{2} - \dfrac{1}{100}} = 2.45\ \Omega$$

$$I_{sh} = 40.8\ \text{mA}$$

$$I = 41.8\ \text{mA}$$

$$R_e = 73.5\ \Omega$$

$$R_r = 73\ \Omega$$

Note that 40 mA overloads most commercial dry cell batteries.

If there are advantages to having such a low value for a midscale reading, a regulated voltage supply can be used or a strong battery, the output impedance of which can be calculated and its value included in R_r.

Adjustment can be provided for R_r and R_{sh}, R_r providing full-scale adjustment and R_{sh} midscale adjustment. (Full scale in this instance is a reading of ∞ ohms.)

If adjustment is provided, a precision resistor must be available for midscale adjustment; otherwise, it will not guarantee a more accurate reading than a circuit built with carefully prepared and pre-measured resistors.

6.1.3 Ohmmeter Loading Effect

A series ohmmeter with a very strong source has no loading error. Any error introduced comes from the unstable value of the source voltage. In contrast, the parallel ohmmeter has a loading effect error.

Consider an ohmmeter such as the one shown in Figure 6.7 with a 1-V source and a 1-mA 100-Ω movement. The reading will be correct at both ends of the scale ($R = 0$ and $R = \infty$): $R_r = 900 \ \Omega$.

At any other point, the reading will be off because R_x introduces a variable load on the source. At half scale, for example when $R_x = 100 \ \Omega$, the resistance of the loop will be 950 Ω. The current

$$I = \frac{V}{R} = \frac{1}{950} = 1.05 \text{ mA}$$

Half of it goes through the meter (0.525 mA). The reading on an uncorrected hyperbolic scale will be 110.5 Ω. A scale can be constructed, though, which *includes* the loading effect, and it is

$$I_m = \frac{V}{R_r + R_m + \dfrac{R_r R_m}{R_x}}$$

(Its derivation is shown in part VI of the Appendix.)

The amount of loading depends on the R_r/R_m ratio, but since R_r depends on I_m, a high meter sensitivity results in a high R_r value, which, in turn, gives a high R_r/R_m ratio and a lower loading error.

6.2 BRIDGES

Series and parallel ohmmeters have their limitations. For accuracy, they depend on a stable, known voltage supply, and the precision of the measurement is limited by the fact that no partial amplification of the scale is possible. The measurement of low resistances, and particularly, small changes in a resistance, cannot be made accurately using these methods.

For measuring electrical components, series or parallel ohmmeters usually are accurate enough. Sensors for process variables (strain gauges and resistance temperature detectors, for example) require accuracy and precision obtainable only by using a bridge.

Measurement of inductance and capacitance can be done using ac bridges, which are discussed briefly in Section 6.2.2.

6.2.1 DC Bridges

In a bridge there are two arms to carry the current. Each arm is a voltage divider. The measurement is taken at midpoint to compare the *voltages* (Figure 6.9). This arrangement solves the problem of the consistency of the voltage supply. Small changes in the supply dc affect both branches equally, not disturbing the *ratio* of the voltages. If the two midpoint voltages are *equal*, the bridge is balanced.

The bridge balance depends on the ratio of the resistances in the two arms, not on their actual value. What is measured is the *resistance change*, which is exactly the type of measurement required by most resistance-change-type sensors and measuring elements. Figure 6.10 shows a few examples of a balanced bridge. In all cases the voltage between A and O equals that of between B and O.

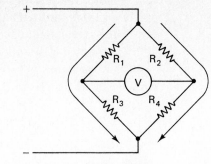

Figure 6.9

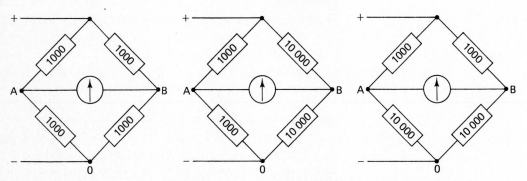

Figure 6.10

If the voltage between A and O is V_x and B and O is V_y, the voltage-divider equation can be written

$$\frac{V}{R_1 + R_3} = \frac{V_x}{V_3}$$

$$V_x = \frac{R_3 V}{R_1 + R_3}$$

(the indices of R as on Figure 6.11), and

$$\frac{V}{R_2 + R_4} = \frac{V_x}{R_4}$$

$$V_x = \frac{R_4 V}{R_2 + R_4}$$

From these equations, the voltage difference between A and B is

$$V_x - V_y = \frac{R_3 V}{R_1 + R_3} - \frac{R_4 V}{R_2 + R_4} = V\left(\frac{R_3}{R_1 + R_3} - \frac{R_4}{R_2 + R_4}\right)$$

which can be a positive or negative value. For the bridge to be in balance:

$$\frac{R_3 V}{R_1 + R_3} = \frac{R_4 V}{R_2 + R_4}$$

which, after being divided by V, multiplied by the denominators, and with $R_3 R_4$ deducted from both sides, gives

$$R_1 R_4 = R_2 R_3$$

the basic Wheatstone bridge equation.

The measurement of the resistance can be done by *balancing the bridge* or *calculating the resistance* from the voltage caused by bridge unbalance between points A and B.

For an *accurate measurement* the following conditions must be met:

1. All resistors must be high-grade stable ones, their value known to the required accuracy.
2. The current through the resistor must not cause heating to the degree that may alter the value of the known or unknown resistors.
3. The detector must have a high resistance to limit the cross-current between A and B.

If there is a less than infinite resistance between A and B, current will flow from A to B if the bridge is unbalanced. This complicates the solution of the circuit, and Thévenin's theorem must be used to find the voltage difference between A and B. This fact dictates that when the unknown resistance is measured by the voltage, the voltmeter must have a very high input impedance. This rules out the use of a PMMC movement, which may also have an insufficient sensitivity. By using a high-input-impedance voltmeter of the required sensitivity, the cross-current induced error can be kept safely below acceptable levels.

6.2.1.1 Measuring Resistance by Bridge Balancing

The simplest, most accurate way of measuring the resistance in a bridge is by bridge balancing. If three resistances are known, the fourth can be calculated when the bridge is balanced. To do that the bridge requires a *null detector* and an *adjustable resistor*.

The adjustable balancing and the measured resistors are in *adjacent arms*. The usual arrangement is shown in Figure 6.11. In this arrangement the resistance of the null detector is immaterial, since when A and B are equipotential points, no current is going to flow through the meter. On the other hand, the null detector must be very sensitive for good reading precision. The null detector should clearly indicate a 1-μA current.

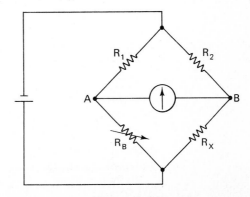

Figure 6.11

The adjustable resistor's setting must be directly readable (otherwise, you wind up with two equal and unknown resistances). The best arrangement is the use of a precision multiturn potentiometer with an easily readable dial or a digital readout attachment. For the sake of simplicity, R_1 and R_2 are equal, and their values are close to the expected value of R_x. As long as $R_1 = R_2$, *the adjusted value of R_B is equal to R_x* when the bridge is balanced.

When measuring *low resistances*, it may be a problem to obtain a multiturn potentiometer of sufficiently low value. In this case R_1 to R_2 ratio can be altered, and when the bridge is balanced, R_B to R_x will be the same ratio.

Example 6.7

A measuring bridge is constructed by $R_1 = 1000\ \Omega$, $R_2 = 20\ \Omega$, and R_B a 10-turn 1000-Ω potentiometer. What is the value of R_x if the bridge balances when R_B, is adjusted to 4.735 turns (assuming that the readout dial has 100 divisions readable to half-divisions)?

If 10 turns = 1000 Ω, then

$$\frac{10\text{ turns}}{1000\ \Omega} = \frac{4.735\text{ turns}}{x\ \Omega}$$

and

$$x = \frac{4.735 \times 1000}{10} = 473.5\ \Omega$$

Since

$$\frac{R_1}{R_2} = \frac{1000}{20} = 50$$

$$\frac{R_b\text{ active}}{R_x} = 50$$

too, and

$$R_x = \frac{473.5}{50} = 9.47\ \Omega$$

The supply voltage is not part of the calculation for a balanced bridge. Its value is limited by the fact that too high a supply voltage heats up the resistors, and too low a supply voltage makes null detection more difficult. A 1-V supply in the example shown will result in a $1/{\sim}30 = 33$ mA current and 33 mW to be dissipated by the resistors, which is a safe value if the components are not miniature.

A 10-V supply, however, will result in a 0.33-A current and a 3.3-W power, which may cause problems. (The current was calculated by dividing the supply voltage with the approximate full resistance of the bridge.)

The layout of the components must be *clean*, the connecting

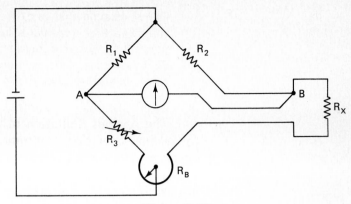

Figure 6.12

wires *oversize*, all connections soldered, and connected by terminal devices or otherwise *positively and firmly joined*, since any unwanted resistance in the circuit nullifies the result.

When measuring *process variables* with a resistance-changing device, the bridge must be balanced first, to find the *initial* value of the sensor, and the final measured resistance compared to that value. Since the resistance change can be small compared to the initial resistance, an arrangement such as that shown in Figure 6.12 is preferred. In the figure R_3 is about the same value as R_x initially. It is adjustable to enable rough prebalancing of the bridge. The value of R_B is selected to be about the same resistance as the expected resistance *change*.

R_x can be any distance from the bridge and the long connecting wires may cause trouble. Their initial effect can be balanced out, but temperature changes (sunshine, for example) may cause enough resistance change to include a serious error, because of the small span of the resistance change measured.

Note that any change of the position of R_B has a *double effect*. It increases the resistance of one arm and decreases the resistance of the other arm at the same time.

Example 6.8

R_x is a strain gauge with a nominal resistance of 200 Ω. R_1, R_2, and R_3 are also selected to be 200 Ω. R_B is a 1-Ω 5-turn potentiometer.

When the bridge is balanced *before* the measurement with R_x, a position of 3.731 turns on R_B balances the bridge (the turns are counted such that zero turns shows zero resistance in the branch of R_x and 1 Ω is the branch of R_3).

When the measurement is made, the bridge is rebalanced by R_B, and its new position is 1.560 turns. The resistance change of R_B is 3.731 – 1.560 = 2.171 turns, that is,

$$\frac{2.171 \times 1}{5} = 0.434 \ \Omega$$

The resistance change of R_x is double that of $0.868\ \Omega$. Although the resistance change is less than 1% of the sensor, it is still possible to obtain a measurement with three- or four-significant-digit accuracy. The accuracy is limited only by the quality of resistors and the sensitivity of the null balance meter.

The obvious limitation of the null balance method is the need for manual operation. Continuous measurement requires a full-time operator.

To automate the reading, a sensitive detector can be employed instead of the null indicator, and the unbalance amplified and fed to a two-phase motor to rebalance the bridge. Some time ago these types of instruments were widely used, with some elegant and ingenious designs. The reason for the use of these masterful but complicated (= expensive) devices was the difficulty of amplifying the unbalance voltage into a practical output voltage or current, which is now done simply with high-input-impedance instrumentation or operational amplifiers. The self-balancing instruments are still in existence but only rarely purchased for new applications.

6.2.1.2 *Measuring Resistance by Unbalanced Voltage*

The equation

$$\Delta V = V\left(\frac{R_3}{R_1 + R_3} - \frac{R_4}{R_2 + R_4}\right)$$

can be solved for any one of the resistors. In practice, the bridge is manually balanced before measurements. It is good precaution to know the resistance of each resistor to the required precision. Since the subtraction of the two terms may eliminate the first significant digit, it may be necessary to know the resistances to an accuracy of five or six significant digits. The bridge itself is likely to be constructed like the one shown in Figure 6.12 where R_3 is the coarse and R_B the fine zero adjustment.

Example 6.9

To measure temperature with a resistance temperature detector (RTD), a bridge is set up with $V = 3$ V; R_1, R_2, and R_3 with $100.00\ \Omega$ each; R_B a 5-Ω 10-turn potentiometer; and R_x with a nominal value of $100\ \Omega$.

Before measurement is made, the bridge is manually balanced by R_B, which happens when it is at 3.92 turns. When the measurement is taken, a very high input impedance voltmeter detector shows 0.0635 V.

At 3.92 turns (out of 10), $1.96\ \Omega$ is added to the sensor (out of 5 Ω), and $5 - 1.96 = 3.04\ \Omega$ added to R_3. The value of R_3 for the calculation is therefore $100 + 3.04 = 103.04\ \Omega$ (and the original value of R_x is, by the way, $103.04 - 1.96 = 101.08\ \Omega$ if R_1 equals R_2).

Substituting these values into the equation, we have

$$0.065 = 3 \left(\frac{100}{200} - \frac{x}{103.04 + x} \right)$$

$$\frac{0.065}{3} = 0.5 - \frac{x}{103.04 + x}$$

$$\frac{0.065}{3} - 0.5 = -0.47833 = -\frac{x}{103.04 + x}$$

$$0.47833x \times 49.287 = x$$

$$x = 94.48 \ \Omega$$

Since x is R_x plus the 1.96-Ω part of R_B, the resistance *change* can be calculated subtracting x from the value of R_3 plus the 3.04-Ω part of R_B:

$$103.04 - 94.48 = 8.56 \ \Omega$$

Employment of high-input-impedance amplifiers made automatic null balancing obsolete. Almost all bridge input instruments manufactured lately amplify the detected voltage for readout or for a transmitted standard signal.

6.2.1.3   Kelvin Bridges

The problem with measuring very low resistances (1 Ω and lower) is that the resistance of the leads and connecting terminals cannot be ignored. (Bridges, like other circuits, always look neat on drawings. The same cannot be said about circuits assembled from components.)

Even when the layout of components is well thought out, the connections are neatly made using suitable wire and hardware, it is likely that the connecting leads will have some resistance, which can be a cause of error for low resistance measurements.

Extra resistance coming from connecting wires and terminals will have no effect on the resistance, though, if the ratio of the resistances in the two arms connecting the detector is the same as the ratio of the adjacent resistances. In Figure 6.13 this is true if the ratio

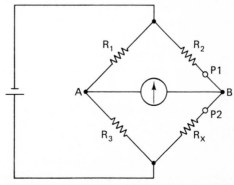

Figure 6.13

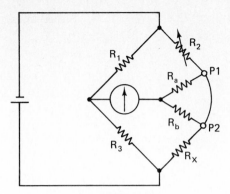

Figure 6.14

of R_2/R_x equals $(P1 - B)/(P2 - B)$, where $P1 - B/P2 - B$ is the ratio of the resistance of wires, and so on, between the points P1 to B and P2 to B. It was Kelvin who first constructed a bridge using this theory and the bridge is named after him.

In practice two extra resistors are employed. The name is given sometimes as the Kelvin double bridge (Figure 6.14). The ratio of the resistors R_a/R_b must be the same as R_1/R_3. When the bridge is balanced, the ratio of R_2/R_x will therefore also be the same.

R_a or R_b can be an adjustable resistor to prebalance or standardize the bridge. R_2 to R_x ratio depends on the expected value of R_x and usually runs from 10/1 to 100/1.

The Kelvin bridge is strictly a laboratory instrument; it requires high-grade precision components.

6.2.2   AC Bridges

A detailed discussion of ac bridges requires some knowledge of ac theory. A brief discussion of the operation, types, and uses of ac bridges can be made, however, based on a basic knowledge of the behavior of circuit components in an ac circuit.

Resistance (pure ohmic resistance) is handled in an ac circuit in the same way as it is in a dc circuit. Its value is independent of the frequency. The opposition to current caused by inductors and capacitors in an ac circuit is called the *inductive and capacitive reactance.* Its value depends on the frequency. Inductive reactance *increases* and capacitive reactance *decreases* as the frequency is *increasing.* (The resistance of a capacitor is infinity and the pure inductor is zero in a dc circuit *except* when the circuit is switched on or off.)

In an ac circuit resistance and reactance cannot be simply added mathematically. The best way to represent their behavior is on a graph (Figure 6.15). (*Note:* Since the reactance values are for a given frequency, they will not be notated in the text as ωL_x or $1/\omega C_x$.) Resistance (R) of a component (or group of components or a circuit) is measured *right* from zero. Inductive reactance (X_L) of a component is measured *up* from zero. Capacitive reactance (X_C) of a component is measured *down* from zero. The net *reactance* of the compo-

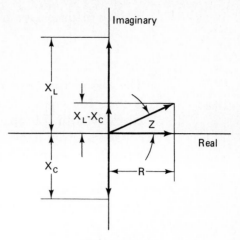

Figure 6.15

nent is found as $X = X_L - X_C$. The *impedance* (Z) of the circuit is found by the *vectorial* summation of X and R. The resultant *phasor* has a magnitude ($\sqrt{X^2 + R^2}$) and a phase angle (tan X/R).

An ac bridge may have any combination of resistive, inductive, or capacitive components, but to balance the bridge the *impedance ratios must be equal*. It means, in practice, that the resistance ratio and the reactance ratio must be established separately. (Capacitors and coils also have some ohmic resistance, which is usually shown separately in ac bridge sketches.)

Ac bridges can be *laboratory type*, used to measure the impedance or reactance of a component, or the supply frequency. These bridges are *balanced* for measurement. Or ac bridges can be used in combination with capacitive or inductive *sensors* to measure process variables, in which case the detected ac is amplified (rectified as necessary) to produce a proportional voltage or current signal.

The simplest ac bridge is the *comparison bridge*. In a comparison bridge the two arms are *identical* when the bridge is balanced (Figure 6.16). Similarly to the balance equation for Wheatstone bridges, for the *resistance* of the components the equation can be written

$$R_1 R_x = R_2 R_s$$

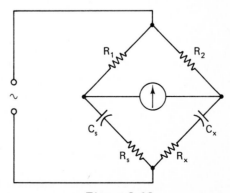

Figure 6.16

and for the *reactance* of the components we can write

$$R_1 \frac{1}{C_x} = R_2 \frac{1}{C_s}$$

The reason for writing $1/C_x$ instead of C_x (and the same with C_s) is that, for a given frequency, the reactance of a capacitor changes *inversely* with the capacitance. R_x and C_x can be calculated from the equations given above. R_1 and R_2 can be equal or be at a certain ratio if the available standard variable resistor (C_s, R_s) is different from the one to be measured (the way it is shown in Figure 6.8).

The measurement of inductances can use a similar bridge, but inductors instead of capacitors, and the reactance equation will be

$$R_1 L_x = R_2 L_s$$

since the reactance of an inductor increases with the inductance. Because of the varying amount or resistance in the capacitors or inductors (high-Q or low-Q coils), specially constructed bridges offer advantages in balancing and component ranging. These are:

Wien bridge	to measure frequency or capacitance
Schering bridge	to measure capacitance
Maxwell bridge	to measure inductance (low-Q coils)
Hay bridge	to measure inductance (high-Q coils)
Owens bridge	to measure inductance

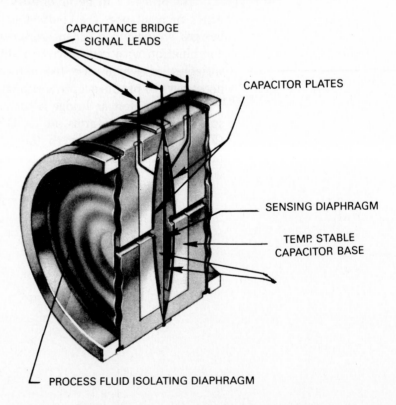

Figure 6.17. © Rosemount, Inc. 1971.

Ac bridges to measure process variables are built like comparison bridges. Supply frequency is selected to give a comfortably measurable reactance change when the measurement is made. Figure 6.17 shows an industrial sensor. The linearity of these measurements is not always assured. The received and processed signal then has to be linearized by a compensating circuit.

6.4 PROBLEMS

1. Calculate R_r and, where necessary, R_{sh} for series-type ohmmeters with the following data.

	Supply Voltage (V)	I_m (mA)	R_m (Ω)	Midscale Resistance (Ω)
(a)	1.5	1.0	100	1000
(b)	3	0.5	250	10 000
(c)	1.5	2	150	5000
(d)	6	0.05	300	200
(e)	4.5	1.0	150	350
(f)	1.5	1.5	100	100

2. Calculate R_r and R_{sh}, where necessary, for the following *parallel* ohmmeters.

	Supply Voltage (V)	I_m (mA)	R_m (Ω)	Midscale Resistance (Ω)
(a)	1.5	0.1	200	200
(b)	1.5	1	100	50
(c)	3	1.5	200	20
(d)	4.5	0.05	500	10

3. Construct and draw a multirange ohmmeter with a 3-V supply and a 0.1-mA 300-Ω movement for 100-, 1000-, 10 000-, and 30 000-Ω midscale resistances.

4. Construct a scale, 10 cm in radius, 90°, for a series ohmmeter with a midpoint resistance of 500 Ω.

5. Calculate the R_4 value that will balance a bridge if the resistors in Figure 6.11 are as follows.

	R_1 (Ω)	R_2 (Ω)	R_3 (Ω)
(a)	1000	2000	1000
(b)	820	100	5600
(c)	113	226	500

6. Find the resistance change of a sensor in a bridge built as in Figure 6.12 based on the following data.

	Potentiometer		Initial Balance	Final Balance
(a)	10 turns	100 Ω	3.27 turns	6.79 turns
(b)	5 turns	10 Ω	0.88 turn	3.05 turns
(c)	10 turns	1 Ω	9.16 turns	5.23 turns

7. Calculate the detected voltage in the following dc bridges (resistor notation as in Figure 6.9).

	R_1 (Ω)	R_2 (Ω)	R_3 (Ω)	R_4 (Ω)	Supply Voltage (V)
(a)	100	100	100	103.15	6
(b)	10 000	10 015	11 613	9987	3
(c)	5000	1000	200	42.9	10
(d)	200	200.3	200	197.04	12

8. Calculate the resistance change of the sensor in the following bridges, which are built like the one in Figure 6.12.

	Supply Voltage (V)	Potentiometer		Balance Setting (turns)	$R_1 = R_2 = R_3$ (Ω nominal)	Detected (mV)
(a)	3	10 turns	1 Ω	2.77	100	7.25
(b)	5	1 turns	2 Ω	0.635	1000	45.0
(c)	10	10 turns	10 Ω	4.91	200	1.07

7

Potentiometric Circuits

7.1 BASIC CONSIDERATIONS

PMMC movements more sensitive than 50 μA f.s.d. are either expensive or vulnerable or both. When measuring EMF sources having a high output impedance, it is necessary to amplify the signal. Amplifiers with very high input impedance (usually differential amplifiers with MOSFET input devices) can safely measure sources such as thermocouples, which have an output impedance of about 20 000 Ω. There are still signal sources with much higher output impedance: for example, pH electrodes and crystals. These can be measured by an amplified signal, as long as the effect of loading is known and the output signal is corrected accordingly. If a high-accuracy measurement is to be made of a weak signal (that is, a source with high output impedance), a *potentiometric circuit* is used. A balanced potentiometric circuit does not require any energy from the source; therefore, it does not load it. The measured EMF is the true open-circuit potential (hence the name).

Before the development of very high input impedance solid-state operational amplifiers, the only way to measure a weak source was with a potentiometric circuit. It shares, nevertheless, one property with balanced bridges: It requires a manual operator, or automatic balancing has to be added.

Automatic balancing is expensive and requires maintenance. It is used now only when no other solution is available. This limits the use of potentiometric circuits in industrial measurements. They are less prevalent now than they were 15 to 20 years ago. Potentiometric circuits still retain their importance in high-accuracy laboratory work.

The theory of potentiometric circuits is based on the fact that current can flow only if there is a voltage (potential) difference; if no potential difference, there is no current. If two sources of exactly equal voltage are connected, one will not discharge into the other, even if their output impedance are vastly different, as can be seen in Figure 7.1.

If a sensitive galvanometer is wired in series into the circuit it will, naturally, show zero. (Zero current *or* zero voltage: When one is

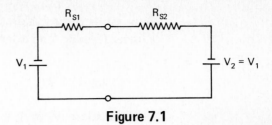

Figure 7.1

zero, the other must be also. Using multimeters for zero detection, the meter must be switched to its *most sensitive* range, voltage or current.) The inclusion of a sensitive galvanometer as a *null detector* verifies the equality of sources when it shows zero current. The potentiometric circuit is a *variable voltage source* which is connected to the source to be measured through a null detector. The variable source is adjusted until the null balance indicates zero. The voltage can now be determined from the variable source. The variable source of a potentiometer is invariably a precision resistor with an adjustable wiper (Figure 7.2). As the adjustable wiper is moved from 0 to V, the potential between 0 and the wiper moves in the same way from 0 to V. If the resistor is a *linear* resistor, as shown in Figure 7.2, the position of the wiper can be measured by its distance from 0 and V. The potential between 0 and the wiper will be

$$V_i = \frac{a}{lV}$$

The interesting thing is that the actual value of the resistor is not part of the equation. Only the *ratio of the resistance* counts.

Example 7.1

What will be the potential (V_i) in the circuit shown in Figure 7.2 if V = 1.875 V, the total length of the resistor (l) = 1.000 m, and a = 0.273 m?

$$V_i = \frac{a}{l} V = \frac{0.273}{1.000} \; 1.875 = 0.512 \text{ V}$$

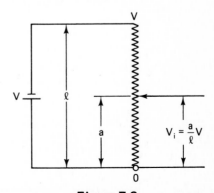

Figure 7.2

It is more comfortable to use a multiturn potentiometer instead of a linear resistor. The calculation is the same.

Example 7.2

Calculate V_i if $V = 1.507$ V and the setting of the 10-turn potentiometer is 4.765 (zero setting is when the wiper is at 0).

$$V_i = 1.507 \, \frac{4.765}{10} = 0.7181 \text{ V}$$

The full potentiometer is shown in Figure 7.3. (The resistor will be shown as linear whether it is linear or rotary.) When the potential to be measured (V_i) is connected, the wiper is adjusted until the null detector detects zero current. Since the voltage between 0 and the wiper must be V_i, this voltage is read by the wiper position.

The accuracy of the measurement depends on two conditions.

1. V must be known as accurately as the reading is to be made. That is, if it is expected to make a voltage measurement to three-digit accuracy, the voltage of the source must be known as well or better.

2. The resistor must be linear, readable to the required precision, and have the necessary resolution. (Resolution in wire-wound resistors depends on the resistance between two turns, which is the minimum increment as the wiper slides from one to the other.)

Although the actual value of the resistor does not come into the calculation, its value must be considered carefully. A *low* value for the resistor will load down the source, and although the initial drop itself will not damage the accuracy of the measurement, a premature exhaustion of the source will cause problems.

A *very high* value for the resistor makes it susceptible to problems with moisture or dirt, which can act as a parallel current path, distorting the linearity of the resistance. If the resistor is used in

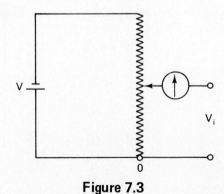

Figure 7.3

laboratory or controlled clean circumstances, resistors up to 100 kΩ can be safely used. For higher resistances, special precautions must be taken.

In most cases, the potentials to be measured are small, much smaller than the supply voltage, V. It is unnecessary to make the whole resistor be an adjustable one when only a small part is to be used regularly. In this case the resistor is made up of two parts: an adjustable resistor, for the required range of potential inputs, and a fixed resistor for the rest (Figure 7.4). Assume that $V = 1.126$ V; the intended input is from a thermocouple which generates about 14 mV, and a 10-turn 1000-Ω multiturn resistor is to be used for R_a.

The resistance of the multiturn potentiometer must be accurately known, or measured. Any error will be transmitted to the measurement. To measure 14 mV comfortably, R_a should drop 25 mV (20, 25, or 30 mV as the measurer decides, but *not* 14 or 15 mV, which may not be enough).

Since the same current goes through both resistors:

$$\frac{V_1}{R_a} = \frac{V - V_1}{R_b}$$

$$R_b = R_a \frac{V - V_1}{V_1} = 1000 \frac{1.126 - 0.025}{0.025} = 44\,040 \ \Omega$$

The loading of the source, by the way, will be

$$I = \frac{V}{R} = \frac{1.126}{44\,040 + 1000} = 25.5 \ \mu A$$

a very safe figure.

Using the arrangement shown in Figure 7.4, the calculation is the same, but both the voltage and the setting refers to the variable resistor only. If, measured with the potentiometer above, the balance is achieved at 6.79 turns, what is the measured potential, V_i?

$$V_i = 25 \ (\text{mV}) \frac{6.79 \ (\text{turns})}{10 \ (\text{turns})} = 17.0 \ \text{mV}$$

The fixed resistor can be a decade box or can be made up using several known values in the proper combination. In either case it is important to have the resistance as accurate as the measurement requires, since the full voltage drop across the adjustable resistor depends on the resistance ratios.

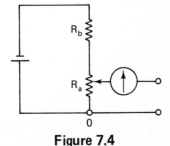

Figure 7.4

7.2   VOLTAGE COMPENSATION

The best supply is a regulated voltage source which is known (or can be measured) to the required accuracy. In many cases a battery (dry cell) is used. The types of batteries that can store enough energy for extended operation (carbon zinc, alkaline, etc.) slowly drop their voltage in use. The types of batteries that have a very accurately known potential are too weak to be used as main voltage supplies. Therefore, potentiometric circuits using regular batteries must have a way of voltage adjustment. To do this, an adjustable resistor is included in series with the main resistor, the potential *across the main resistor* is arbitrarily fixed to a value, and the adjustable resistor is used initially to get this value, and readjusted as the main voltage supply weakens, to maintain this value.

The *adjusted* voltage across the main resistor can be measured, but the customary arrangement is to include a known precise voltage supply parallel to the main supply with a switch and a null detector. The null detection shows when the calibrating power supply is switched on that the voltage coming from the main source is equal to that of the calibrating source.

Figure 7.5 shows the arrangement. The total voltage drop across the main resistor is the same as the potential of the calibrating supply V_c. Naturally, V must be somewhat higher than V_c.

When the switch is closed, R_c is adjusted until the detector shows null, showing that the voltage between A and 0 is exactly equal to V_c. From time to time the switch is closed, and if the null detector deviates, R_c is readjusted until there is balance. It is important to remember that for calculations, the voltage drop between A and 0 rather than the value of the supply V should be used.

The reference unit used for the voltage adjustment (V_c) can be a *standard cell* such as the saturated or unsaturated Weston (cadmium) cell or a constant-voltage circuit using a zener diode.

The Weston cell (Figure 7.6) has a potential of 1.0186 V at 20°C which changes less than 1 μV per year, but it changes with the temperature at the rate of 40 μV/°C. The unsaturated Weston cell has a negligible temperature change, but its output drops about 100 μV per year. Weston cells should be very carefully handled. Their use is limited mainly to the laboratory these days.

There are very ingenious elaborate circuits for automatic volt-

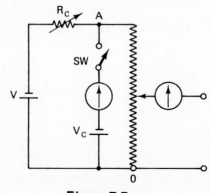

Figure 7.5

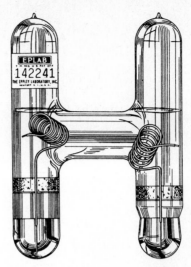

Figure 7.6 (Courtesy The Eppley Laboratory, Inc.)

age compensation. The use of zener diode stabilized voltage sources outdated most of these devices. There is, however, the possibility that a reference voltage source is used, but for some reason it is inconvenient to select the voltage drop across the main resistor to be the same value.

If the reference voltage is *less* than the selected voltage drop across the main resistor, the arrangement shown in Figure 7.7 can be used. The circuit of V_c is connected to the main resistor at the point where the voltage is equal to the reference voltage. That is, the voltage between C and 0 equals V_c. The location of this point is calculated in the same way as that of the wiper position for the input.

Example 7.3

The resistor is a 80-cm 20-kΩ resistor with 2.5 V across it. V_c is a Weston cell (1.019 V).

The point to join the reference arm is calculated as follows:

$$a = l \, \frac{V_c}{V} = 80 \, \frac{1.019}{2.500} = 32.61 \text{ cm}$$

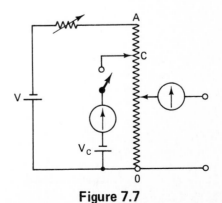

Figure 7.7

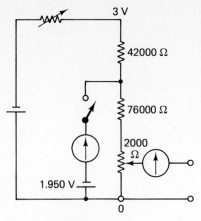

Figure 7.8

The calculation is only slightly more complicated if a variable and fixed resistor combination is used. Since it is impossible to tap into a fixed resistor, two fixed resistors must be used, as shown in Figure 7.8.

Example 7.4

Let the adjusted voltage be 3 V, $V_c = 1.950$ V, and a 2-kΩ adjustable resistor is used to measure 50 mV. There will be two equations. To find the resistor between C and the variable resistor:

$$\frac{50 \ (\text{mV})}{2 \ (\text{k}\Omega)} = \frac{1950 \ (\text{mV}) - 50 \ (\text{mV})}{x \ \text{k}\Omega}$$

$$x = \frac{2 \times 1900}{50} = 76.00 \ \text{k}\Omega$$

and for the other resistor (one of several possible equations)

$$\frac{1.950 \ (\text{V})}{78 \ \text{k}\Omega} = \frac{3.000 - 1.950 \ (\text{V})}{x \ \text{k}\Omega}$$

$$x = \frac{78 \times 1.050}{1.950} = 42.00 \ \text{k}\Omega$$

The reference voltage can be higher than the voltage across the main resistor. In this case a resistor must be placed in series between the reference cell and the main resistor to drop the voltage across the main resistor to exactly the selected value. Figure 7.9 shows the arrangement.

Example 7.5

To calculate R_c, the value of the main resistor must be accurately known. The same type of equation is used here again.

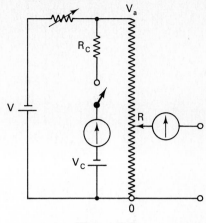

Figure 7.9

Since

$$\frac{V_a}{R} = \frac{V_c - V_a}{R_c}$$

$$R_c = R \, \frac{V_c - V_a}{V_a}$$

For example, if $V_c = 4.508$ V, $V_a = 3.000$ V, and $R = 15\,000\ \Omega$,

$$R_c = 15\,000 \, \frac{4.508 - 3}{3} = 7540\ \Omega$$

Note in the equation above that V_a is the voltage across R. The adjustable resistor sets V_a to its desired value.

Automatic null balancing for potentiometric circuits is still used, although there is a minimum voltage inbalance which is necessary to initiate correction, which minimum may be enough to introduce a loading, and with that, an error. Once the idea of complete balance, which guarantees no loading error, is given up, the amplification of the voltage signal using a very high input impedance amplifier is simpler.

7.3 PROBLEMS 1. Calculate the voltage between the wiper and common (0) for the following potentiometric circuits.

	Supply (V)	*Resistor*	*Adjustment*
(a)	1.50	100.0 cm	7.65 cm
(b)	1.875	10 turns	7.092 turns
(c)	3.150	35.00 cm	26.73 cm
(d)	10.0	5 turns	3.39 turns

2. Calculate the value of the fixed resistor (R_b) for the following potentiometric circuits.

	Voltage Supply (V)	Intended Measurement Range (mV)	Adjustable Resistor (*turns*)	Resistance (Ω)
(a)	1.215	0–100	10	100
(b)	1.864	0–7.5	1	1000
(c)	2.500	0–15	5	5000
(d)	2.84	0–100	10	1000

3. Calculate the measured voltage with the potentiometers given in Problem 2 if the measuring potentiometers are adjusted for zero at:

   (a) 6.951 turns

   (b) 0.613 turn

   (c) 4.075 turns

   (d) 3.65 turns

4. Calculate the location of the joining point (c) between the reference circuit and the main resistor if:

	Voltage Across R (V)	V_c	Length of Resistor (cm)
(a)	2.000	1.019	100
(b)	1.650	0.750	40

5. Calculate the two fixed resistors to accommodate the branch to the reference circuit for the following potentiometric circuits.

	Voltage Across R (V)	V_c	Measured Range Across V_a (mV)	Multiturn Resistor (*turns*)	Resistance (Ω)
(a)	1.500	1.019	10	10	1000
(b)	2.333	2.000	25	10	200
(c)	4.50	3.618	5	20	500
(d)	9.000	4.500	150	1	100

6. Calculate the voltage-dropping resistor R_c to accommodate the reference voltage source in the following potentiometric circuits.

	Voltage Across R (V)	V_c (V)	R (Ω)
(a)	1.000	1.019	20 012
(b)	1.673	2.500	6750
(c)	2.500	3.125	45 200
(d)	3.35	5.500	29 971

Appendix

I. Definitions

Mole: the amount of substance of a system which contains as many elementary entities as there are atoms in 0.012 kg of carbon 12. When the mole is used, the elementary entities must be specified and may be atoms, molecules, ions, electrons, other particles, or specified groups of such particles.

Candela: the luminous intensity, in the perpendicular direction, of a surface of 1/600 000 m^2 of a black body at the temperature of freezing platinum under a pressure of 101 325 Pa.

II. Derived SI Units without Special Names

Quantity	Description	Expressed in Terms of Other SI Units	Expressed in Terms of Base and Supplementary Units
Area	square meter	m^2	m^2
Volume	cubic meter	m^3	m^3
Speed			
Linear	meter per second	m/s	m $\cdot$ s^{-1}
Angular	radian per second	rad/s	rad $\cdot$ s^{-1}
Acceleration			
Linear	meter per second squared	m/s^2	m $\cdot$ s^{-2}
Angular	radian per second squared	rad/s^2	rad $\cdot$ s^{-2}
Wave number	1 per meter	m^{-1}	m^{-1}
Density, mass density	kilogram per cubic meter	kg/m^3	kg $\cdot$ m^{-3}
Concentration (of amount of substance)	mole per cubic meter	mol/m^3	mol $\cdot$ m^{-3}
Specific volume	cubic meter per kilogram	m^3/kg	m^3 $\cdot$ kg^{-1}
Luminance	candela per square meter	cd/m^2	cd $\cdot$ m^{-2}
Dynamic viscosity	pascal second	Pa $\cdot$ s	m^{-1} $\cdot$ kg $\cdot$ s^{-1}
Moment of force	newton meter	N $\cdot$ m	m^2 $\cdot$ kg $\cdot$ s^{-2}
Surface tension	newton per meter	N/m	kg $\cdot$ s^{-2}
Heat flux density, irradiance	watt per square meter	W/m^2	kg $\cdot$ s^{-3}
Heat capacity, entropy	joule per kelvin	J/K	m^2 $\cdot$ kg $\cdot$ s^{-2} $\cdot$ K^{-1}
Specific heat capacity, specific entropy	joule per kilogram kelvin	J/(kg $\cdot$ K)	m^2 $\cdot$ s^{-2} $\cdot$ K^{-1}
Specific energy	joule per kilogram	J/kg	m^2 $\cdot$ s^{-2}
Thermal conductivity	watt per meter kelvin	W/(m $\cdot$ K)	m $\cdot$ kg $\cdot$ s^{-3} $\cdot$ K^{-1}
Energy density	joule per cubic meter	J/m^3	m^{-1} $\cdot$ kg $\cdot$ s^{-2}
Electric field strength	volt per meter	V/m	m $\cdot$ kg $\cdot$ s^{-3} $\cdot$ A^{-1}
Electric charge density	coulomb per cubic meter	C/m^3	m^{-3} $\cdot$ s $\cdot$ A
Surface density of charge, flux density	coulomb per square meter	C/m^2	m^{-2} $\cdot$ s $\cdot$ A
Permittivity	farad per meter	F/m	m^{-3} $\cdot$ kg^{-1} $\cdot$ s^4 $\cdot$ A^2
Current density	ampere per square meter	A/m^2	A $\cdot$ m^{-2}
Magnetic field strength	ampere per meter	A/m	A $\cdot$ m^{-1}
Permeability	henry per meter	H/m	m $\cdot$ kg $\cdot$ s^{-2} $\cdot$ A^{-2}
Molar energy	joule per mole	J/mol	m^2 $\cdot$ kg $\cdot$ s^{-2} $\cdot$ K^{-1} $\cdot$ mol^{-1}
Molar entropy, molar heat capacity	joule per mole kelvin	J/(mol $\cdot$ K)	m^2 $\cdot$ kg $\cdot$ s^{-2} $\cdot$ K^{-1} $\cdot$ mol^{-1}
Radiant intensity	watt per steradian	W/sr	m^2 $\cdot$ kg $\cdot$ s^{-3} $\cdot$ sr^{-1}

III.   Units Permitted for Use with the SI

Name	Symbol	Definition
Nautical mile	—	1 nautical mile = 1852 m
Knot	kn	1 nautical mile per hour = (1852/3600) m/s
Millibar	mbar	1 mbar = 100 Pa
Standard atmosphere	atm	1 atm = 101.325 kPa
Gal	gal	1 gal = 1 cm/s^2
Roentgen	R	1 R = 2.58 $\times$ 10^{-4} C/kg

IV.   Units That Should Not Be Used with the SI

Quantity	Name	Symbol	Definition
Length	angstrom	Å	1 Å = 0.1 nm
	micron	μ	1 μ = 1 μm
	fermi	fm	1 fermi = 1 femtometer = 1 fm
	X unit	—	1 X unit = 100.2 fm
Area	are	a	1 a = 100 m^2
	barn	b	1 b = 100 fm^2
Volume	stere	st	1 st = 1 m^3
	lambda	λ	1 λ = 1 μl = 1 mm^3
Mass	metric carat	—	1 metric carat = 200 mg
	gamma	γ	1 γ = 1 μg
Force	kilogram-force	kgf	1 kgf = 9.806 65 N
	kilopond	kp	1 kp = 9.806 65 N
	dyne	dyn	1 dyn = 10 μN
Pressure	torr	torr	1 torr = $\dfrac{101\,325}{760}$ Pa
Energy	calorie	cal	1 cal = 4.1868 J
	erg	erg	1 erg = 0.1 μj
Viscosity			
Dynamic	poise	P	1 P = 1 dyn $\cdot$ s/cm^2 = 0.1 Pa $\cdot$ s
Kinematic	stoke	St	1 St = 1 cm^2/s
Conductance	mho	mho	1 mho = 1 S
Magnetic field strength	oersted	Oe	1 Oe $\hat{=}$ $\dfrac{1000}{4\pi}$ A/m
Magnetic flux	maxwell	Mx	1 Mx $\hat{=}$ 0.01 μWb
Magnetic flux density	gauss	G	1 G $\hat{=}$ 0.1 mT
Magnetic induction	gamma	γ	1 γ $\hat{=}$ 1 nT
Illuminance	phot	ph	1 ph = 10 klx
Luminance	stilb	sb	1 sb = 1 cd/cm^2
Activity (radioactive)	curie	Ci	1 Ci = 37 GBq
Absorbed dose of ionizing radiation	rad	rad	1 rad = 10 mGy

V.   Example of a Heading in a Calibration Notebook

July 5, 1975, 3 p.m., Measurement Lab. (E 225), Beau Company

Calibrator: John Doe, CIT

Pressure indicator: Acme model no. P-3574, serial no. S390451, input 3–15 lb/in^2

Calibration setup P1-1: Precision pressure gauge, W.T. model no. 312, serial no. 64-198, 0–30 lb/in².

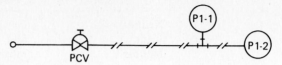

Serial no. 67-2354

P1-2: The calibrated instrument

Room temperature: 72°F

Barometric pressure: 29.72 in. Hg

The instrument was calibrated in a vertical position. On field use protection should be provided to conform to ISA-2B housing standards.

VI.   Derivation of the Parallel Ohmmeter Scale Equation

(Notations as in Section 6.1.2; V_c is the voltage drop across R_x and the movement, R_c, is the combined resistance of R_x and R_m.) The current through the meter

$$I_m = \frac{V_c}{R_m}$$

The current through the loop

$$I = \frac{V_c}{R_c} = \frac{V}{R_c + R_r}$$

From this equation

$$V_c = \frac{VR_c}{R_c + R_r}$$

With the first equation

$$I_m = \frac{VR_c}{R_m(R_c + R_r)}$$

Since

$$R_c = \frac{R_m R_x}{R_m + R_x}$$

$$I_m = \frac{\dfrac{R_m R_x V}{R_m + R_x}}{R_r R_m + \dfrac{R_m R_x}{R_m + R_x}}$$

Simplified by R_m and multiplied (numerator and denominator) by $R_m + R_x$, we have

$$I_m = \frac{VR_x}{R_r R_m + R_r R_x + R_m R_x}$$

or

$$I_m = \frac{V}{R_r + R_m + \dfrac{R_r R_m}{R_x}}$$

where I_m is the meter indication with R_x attached. If I_m is not a convenient value (1 mA, for example) the equation can be expressed as a percentage:

$$S\,(\%) = \frac{100}{I_m}\ \frac{V}{R_r + R_m + \dfrac{R_r R_m}{R_x}}$$

where I_m now is the meter full-scale deflection.

If I_{sh} is used parallel with the meter movement to set the half-scale resistance, the meter current can be calculated from the following equation:

$$I_i = \frac{V}{R_i + \dfrac{R_r R_i}{R_c}} \quad \text{where} \quad R_c = \frac{1}{\dfrac{1}{R_{sh}} + \dfrac{1}{R_i} + \dfrac{1}{R_x}}$$

VII.   Meter Resistance Calculation

The resistance of the loop with R_a in position will be

$$R = R_r + \frac{R_m R_a}{R_m + R_a}$$

The current in the loop

$$I = \frac{V}{R} = \frac{V}{R_r + \dfrac{R_m R_a}{R_m + R_a}}$$

If R_a is adjusted for exactly half the pointer deflection ($I_m/2$),

$$\frac{I_m}{2}\,R_m = R_a I_a$$

but as $I_a = I - I_m$,

$$R_a = R_m\,\frac{I_m}{2I - I_m}$$

Substituting I, we have

$$R_a = \frac{R_m I_m}{\dfrac{2V}{R_r + \dfrac{R_m R_a}{R_m + R_a}} - I_m}$$

This equation can be rearranged to express R_a or R_m :

$$R_a = \frac{R_r R_m I_m}{2V - R_r I_m - R_m I_m}$$

$$R_m = \frac{R_a(2V - R_r I_m)}{I_m(R_r + R_a)}$$

Solutions to Selected Problems

Chapter 1

1. (a) 3,   (b) 1,   (c) 2,
       (d) 5,   (e) 3,   (f) 8
2. (a) 1.757 m^2
3. (a) 160 km
   (b) $3.10 \ \mu\text{m}$
   (c) 0.315 mg
   (d) 4.275 Tg
4. The kilogram
5. (a) 0.40%
6. (a) 16.6 kPa–16.0 kPa
7. (a) 31.16 km
   (b) 5.00 kPa
8. (a) 89.3 kPa–87.3 kPa
   (b) $135.45°\text{C}$–$135.05°\text{C}$ (the rounding of 0.05 is optional)

Chapter 3

1. (a) $9800 \ \Omega$
   (b) $0.0 \ \Omega$
2. (a) $6000 \ \Omega$, $533.33 \ \Omega$, $15 \ \Omega$,
       $30 \ \Omega$ adjustable
   (b) $1000 \ \Omega$, $500 \ \Omega$, $300 \ \Omega$, $25 \ \Omega$,
       $50 \ \Omega$ adjustable
3. (a) 0.89 V (1.72%)
   (b) 0.003 (0.65)
4.

	$R \ \Omega$	V
(a)	5000	2.135
(b)	10 000	0.694

5. (a) 0.57 V   2.5%
   (b) 0.0065 V   4.6%

6.

	V	$R_S \ \Omega$
(a)	6.86	~100
(b)	2.81	~110 O.U.

Chapter 4

1. (a) 477
   (b) 6450
2. (a) $40 \ 650 \ \Omega$, $2500 \ \Omega$,
       $1000 \ \Omega$ adjustable
   (b) $6745 \ \Omega$, $1355 \ \Omega$, $300 \ \Omega$,
       $500 \ \Omega$ adjustable

Chapter 5

1., 3.

	Selected Range	$I_{sh} \ \Omega$	Error %
(a)	0.4 mA	35.7	1.5
(b)	0.5 mA	0.0	0.93
(c)	100 mA	1.01	0.25

2. (a) $0.160 \ \Omega$, $0.0200 \ \Omega$, $0.0200 \ \Omega$
   (b) $0.0750 \ \Omega$, $0.0200 \ \Omega$, $0.00250 \ \Omega$,
       $0.00250 \ \Omega$
4. 2 mA 150   or   1 mA 100
5. (a) $0.497 \ \Omega$
   (b) $158 \ \text{m}\Omega$

Chapter 6

1.

	$R_{sh} \ \Omega$	$R_r \ \Omega$
(a)	200	933.3
(b)	—	9750
(c)	300	100

2.

	R_{sh} Ω	R_r Ω
(a)	0.0	14 800
(b)	100	1450

5. (a) 2000 Ω
   (b) 682.9 Ω

6. (a) 70.4 Ω decrease

7. (a) -0.0465 V
   (b) 0.114 V

8. (a) 0.978 Ω

Chapter 7

1. (a) 0.115 V
   (b) 1.330 V

2. (a) 1115 Ω
   (b) 247.5 kΩ

3. (a) 69.51 mV
   (b) 4.598 mV

4. (a) 50.95 cm

5. (a) 48.1 kΩ 100.9 kΩ
   (b) 2664 Ω 15.8 kΩ

6. (a) 380.2 Ω
   (b) 3337 Ω